SpringerBriefs in Philosophy

SpringerBriefs present concise summaries of cutting-edge research and practical applications across a wide spectrum of fields. Featuring compact volumes of 50 to 125 pages, the series covers a range of content from professional to academic. Typical topics might include:

- A timely report of state-of-the art analytical techniques
- A bridge between new research results, as published in journal articles, and a contextual literature review
- A snapshot of a hot or emerging topic
- An in-depth case study or clinical example
- A presentation of core concepts that students must understand in order to make independent contributions

SpringerBriefs in Philosophy cover a broad range of philosophical fields including: Philosophy of Science, Logic, Non-Western Thinking and Western Philosophy. We also consider biographies, full or partial, of key thinkers and pioneers.

SpringerBriefs are characterized by fast, global electronic dissemination, standard publishing contracts, standardized manuscript preparation and formatting guidelines, and expedited production schedules. Both solicited and unsolicited manuscripts are considered for publication in the SpringerBriefs in Philosophy series. Potential authors are warmly invited to complete and submit the Briefs Author Proposal form. All projects will be submitted to editorial review by external advisors.

SpringerBriefs are characterized by expedited production schedules with the aim for publication 8 to 12 weeks after acceptance and fast, global electronic dissemination through our online platform SpringerLink. The standard concise author contracts guarantee that

- an individual ISBN is assigned to each manuscript
- each manuscript is copyrighted in the name of the author
- the author retains the right to post the pre-publication version on his/her website or that of his/her institution.

Joseph Ganem

Understanding the Impact of Machine Learning on Labor and Education

A Time-Dependent Turing Test

 Springer

Joseph Ganem
Department of Physics
Loyola University Maryland
Baltimore, MD, USA

ISSN 2211-4548   ISSN 2211-4556 (electronic)
SpringerBriefs in Philosophy
ISBN 978-3-031-31003-4   ISBN 978-3-031-31004-1 (eBook)
https://doi.org/10.1007/978-3-031-31004-1

This Springer imprint is published by the registered company Springer Nature Switzerland AG
The registered company address is: Gewerbestrasse 11, 6330 Cham, Switzerland

Preface

A Paradox: Chess Versus Poker

I played a lot of chess when I attended college in the late 1970s. I never attained master status, but I became a reasonably competent and experienced tournament player—skilled enough to win some substantial cash prizes. A campus administrator, who I befriended, knew of my interest in chess. One day he told me that his mother had become fascinated with the game. She had purchased one of the first commercially available chess computers and enjoyed playing against it. However, she had never won a game and wondered if beating a computer at chess was even possible. I told him that yes, computers can be beaten at chess and that I could beat a computer. My friend relayed this to his mother who was intrigued.

Obviously, I could never make such a boast in the present. The 2020's era computer that I write these words on has an inexpensive chess app that can beat the World Chess Champion. But in the 1970s, when the first commercially available chess computers entered the market, computer technology—both hardware and software—was primitive compared to the present. Even though at that point in time, I had never seen a chess computer, let alone played one, I could confidently make that boast because I knew of their existence and their shortcomings. Computers of that era would always beat casual players—like my friend's mother—because of their speedy and infallible ability for extensive and deep calculations. But the computers could not plan and if you want to be reasonably competent at chess, even amateur-level tournament play requires the ability to plan. I cannot calculate all possibilities 10 to 15 moves ahead, but chess has so many possibilities that computers cannot calculate all possibilities that far ahead either. However, I can *plan* 10 to 15 moves ahead, something the computers of that era could not do. Without the ability to plan, computers would randomly meander into lost positions that no amount of calculation could save.

My friend's mother wanted to see her chess computer lose a game. Arrangements were made for me to visit with his mother for an evening. She would make me a home-cooked meal—a real treat when you're eating campus cafeteria food every day—and I would play against her chess computer while she watched. I enjoyed a delicious meal with his mother, a vibrant woman in her 70s, who turned out to be a delightful dinner companion. Afterward, we sat in her living room for coffee and dessert while I took on her computer. As I suspected, I won rather quickly and easily. She was elated as the "I lose" LED on the machine lit up for the first time.

Of course, as a reasonably skilled amateur chess player, I can easily beat humans with a casual knowledge of the game, even though just about all humans can plan. The reason is that planning requires a knowledgebase and the more knowledge you have the better the plan. My chess knowledgebase is so large compared to casual players; I hardly need to reason out problems when I play against them. I can use knowledge recall almost entirely to win without much need for reasoning. Likewise, I stand no chance in chess against master-level players for the same reason. Their knowledgebases are so large that they tower over mine, and they can beat me easily using mostly recall, again with relatively little reasoning needed on their part. Chess only becomes a contest with an uncertain outcome when the two players bring comparable knowledgebases to the chessboard and are forced to use their reasoning abilities within those equitable knowledge frameworks. Reasoning is more difficult and more fallible in comparison with recall, hence the uncertainty in the outcome of the game.

Another game with which I have extensive experience and knowledge is poker. Like chess, my knowledgebase/skillset is far from professional level, although it is much greater than a casual player's knowledgebase/skillset. But, unlike chess, I can never predict the outcome of an evening of poker. Because of the statistical nature of the game, outcomes of individual hands in poker are always uncertain. This means that for relatively small sample sizes—say a hundred hands or so that might be played in an evening—the outcome cannot be predicted. If my friend's mother had owned a poker-playing computer, I could never assert that I could come over for an evening and beat it. But poker is a game of skill, and over the course of thousands of hands, the better players do win. Over the course of tens of thousands of hands extending over many years, I have won more money than I have lost playing poker. On a per hour basis, it's not enough to ever consider quitting my day job, but I clearly haven't been "lucky" for that long a time and with that large a sample of hands. Also, there are much better players than me who can and do make their livings playing poker. Given that poker is a zero-sum game, the winnings must come from the less-skilled players.

As in chess, skilled poker players use both recall and reasoning. A large poker knowledgebase from which to instantly recall facts is an asset. But poker players must always use reasoning because just about every decision is made under a unique set of circumstances. Professional poker players cannot simply overpower their weaker opponents with their substantial knowledge and experience in the way that professional chess players can overpower theirs.

The difference between these two games became apparent as computer technology developed. The early chess computers might have been weak in comparison with an amateur tournament player, but they were instructional. If you could learn enough about chess to beat one, you were on your way to becoming a decent chess player. For poker, beyond learning the mechanics of the game, the early software simulations were not all that useful. Consistently beating a poker program from the late twentieth century predicted nothing about play against actual humans.

In 1997, nearly 20 years after my winning encounter against an early chess computer, a computer beat the World Chess Champion for the first time. But it would be another 20 years, not until 2017, before a computer would be able to beat professional poker players. In terms of hardware and software development, 20 years is eons of evolution. It took many orders of magnitude of additional computing power to move from professional-level chess to professional-level poker, even though at a superficial level, chess appears to be the more complicated game.

Why?

To me this is a paradox that must provide some insight on the future of artificial intelligence (AI), and specifically, insights on how humans and AIs will interact in the future.

The Element of Time: Recall Versus Reasoning

This book arose from trying to understand this paradox. Along the way, I arrived at insights about the impact of AI on twenty-first century work and education that I believe are important enough to share. The many AI researchers involved in developing machines that can play games such as chess, poker, bridge, Go, and others, want to do much more than produce commercial home entertainment products. They want to understand human–machine interactions so that AI can be developed for economically important problems in the workplace. These kinds of strategy games, although played primarily for entertainment, also provide a useful laboratory for generating insights about AI with far-reaching consequences.

As I reflected on the difference between chess and poker, I realized it involved the element of time. Consider the spectacle of the simultaneous exhibition in chess, in which a grandmaster will play dozens of games simultaneously against weaker opponents, typically winning almost all the games. This requires the grandmaster to quickly circulate around a room and make nearly instantaneous decisions at each board. As impressive as this feat appears, for a grandmaster it is not that difficult to pull off. He or she can play strong moves by glancing at only the position on the board and rely on fast recall of extensive chess knowledge. There is no need for any knowledge of either the opponent, or the narrative structure of the game. There is no temporal context to the decision tree.

No such analog to the simultaneous exhibition in chess exists in poker. The decision tree in poker exists within a temporal context. There is no "position" in poker like there is in chess. Decisions are based on the actions the other players have taken in the past and the actions that the other players are expected to take in the future. Strong poker players must act within this temporal framework and pull off the seemingly impossible task of making rational decisions while also being unpredictable. This goal appears oxymoronic. Are not rational decisions, by definition, predictable?

However, over the past century an entire branch of mathematics—Game Theory—has been developed that essentially solves this problem.

But calculations of game-theory-optimal (GTO) play in poker are computationally, extremely intensive. Counterintuitively, the computational resources required for GTO poker are much larger than needed for chess. This is the reason that computers that play professional-level poker were 20 years behind the computers that play professional-level chess. But GTO play is mostly relevant at the highest levels of poker competition because it is difficult for humans to implement perfect GTO play. Mathematically, perfect GTO play insures that over time, a player will never lose and will always win against any other player deviating from perfect GTO play. If all players use perfect GTO play, no one wins. Again, poker is a zero-sum game.

The outputs from GTO solvers provide useful insights and now inform modern poker play. However, average poker players can be successful against average competition by observing their opponents' tendencies and devising ways to exploit them, while at the same time being self-aware of their own tendencies and defending against their opponents' exploits. No computationally intensive GTO solver is necessary. Also, because GTO play is intentionally blind to the human element of poker, paradoxically, game-theory-optimal can be less optimal than a strategy that uses the additional information available about human decision-making that only other humans can discern. But that discernment is a learning process that unfolds in time.

It is for this reason that the early poker programs were not very good. Without the computational resources to devise a mathematically perfect strategy that could not be exploited, these programs had to employ some other reasonable strategy. But observant humans could always discern the strategy and exploit it. In addition, the programs struggled to model the behavior of the humans and devise exploits of their own. In essence, exploitive poker is an exercise in real-time learning and humans were much better at learning than the machines of that era. This may now be changing as AI enters an era in which machines can learn, which will be an important theme in this book.

With these considerations in mind, I began framing my thoughts in terms of the well-known Turing test. But I found that the Turing test was missing the element I now considered most important—time. The Turing test, which involves a human interrogating a machine, presumes both the human and the machine have pre-existing knowledgebases because otherwise, it is not possible to formulate meaningful questions. Where did those knowledgebases come from? The answer is that they had to be learned, which is a process that takes time. While both processes involve thinking, learning is different than knowing, a distinction that the Turing test does not make.

Expertise in chess, poker, or any other field is founded on knowing. People who "know" can access important information quickly and efficiently by recalling what they have already learned. Experts often take years to learn what they know, and in addition, what they learned might have taken other people years to figure out. I am a physicist, and it took me years to learn quantum mechanics, which is a theory about nature that took many other people decades to reason out. But I now have instant access to that extensive knowledgebase—just through recall.

However, learning is a different kind of intelligence. It uses reasoning, which is slow and unsteady, and therefore has timescales that can differ by many orders of magnitude depending on the learning goals and background of the person doing the learning. Reasoning is not instantaneous, and it is much more fallible. Therefore, most people, including myself, are not successful at learning everything that they set out to learn.

We recall instantly what we know; we reason over time to learn. But learning must come before knowing. Machines have always had the ability for fast and efficient recall—superior to the recall ability of humans. But what they recalled had to be facts that were already known. In the past, for machines to provide expertise in chess, or any other area of knowledge, the machines had to be programmed with existing knowledge. But in recent decades, machines are acquiring the ability to learn. This adds a new dimension to AI and its ongoing impact on society.

With this insight, I began my quest to understand how machine learning would impact the future of education and labor. I realized that in this quest it would be useful to enhance the Turing test to include the element of time. Then I would have a useful framework for understanding the novel issues resulting from machines being able to learn.

Baltimore, MD, USA   Joseph Ganem
February 2023

Acknowledgments I thank Francis Cunningham, my colleague at Loyola University Maryland, for helpful feedback, discussions, and suggestions in the preparation of this book. He has also been a valued friend, mentor, and role model throughout my 29 years and counting on the Loyola faculty.

I thank my wife Sharon Baldwin for her 40 years and counting of support for my research and writing, and patience as I continually explore new ideas.

Contents

About the Author

Joseph Ganem, Ph.D. is a professor of physics at Loyola University Maryland. He is an author of numerous scientific papers in the fields of optical materials, lasers, and magnetic resonance and has received grants from Research Corporation, the Petroleum Research Fund, and the National Science Foundation for his research on solid-state laser materials. He has taught physics in the classroom for more than 25 years and has served on the Maryland State Advisory Council for Gifted and Talented Education.

Dr. Ganem is the author of the award-winning book *The Two Headed Quarter: How to See Through Deceptive Numbers and Save Money on Everything You Buy.* He speaks and writes frequently on science, consumer, and education issues and has been a contributor of articles on these topics to the *Baltimore Sun* newspaper. For its 2017 "Best of Baltimore" awards, *Baltimore Magazine* named him one the "Best Baltimoreans" in its people in the media section for the category "Best Defense of Science." In 2018, Springer published his book *The Robot Factory: Pseudoscience in Education and Its Threat to American Democracy.*

Dr. Ganem earned a Ph.D. from Washington University in Saint Louis, a M.S. from the University of Wisconsin-Madison, and a B.S. from the University of Rochester. He did postdoctoral work at the University of Georgia-Athens and at the United States Naval Research Laboratory in Washington, DC.

For more information, visit www.JosephGanem.com.

List of Figures

List of Tables

Chapter 1
Introduction: The Difference Between Knowing and Learning

Abstract Artificial Intelligence (AI) has entered an era in which machines do more than perform activities that in pre-computer times required human intelligence. Machines now learn and adapt to new situations without human intervention. That machines can learn, and even plan, and not need to rely on human programmers for all their actions promises to be more economically disruptive than the introduction of machines that merely performed work associated with intelligence. The difference is that while machines and humans exhibit behaviors that are said to require "intelligence," humans have always had the ability to go a step further. Humans can change and adapt their thinking in response to external feedback as time progresses. As a result, machine learning will fundamentally alter the relationship between humans and machines. Rather than productivity enhancement tools, machines will become, in a sense, co-workers and a division of labor will need to be negotiated. This book argues that "comparative learning advantages" between humans and learning-enabled machines will determine the division of labor. As a result, the future economy will be one in which human capital is valued for time-dependent-learning rather than static expertise that schools have traditionally instilled.

Keywords Machine learning · Philosophy of artificial intelligence · Artificial Intelligence in the workplace · Division of labor between humans and machines · Human capital

Artificial Intelligence (AI) has entered an era in which machines do more than perform activities that in pre-computer times required human intelligence. Machines now learn and adapt to new situations without human intervention. "Learning algorithms" enable machines to modify their actions based on real-world experiences. It is no longer necessary for human programmers to anticipate and provide step-by-step instructions for every possible situation a machine might encounter. Powerful new learning algorithms acting on large data sets enable machines to develop the ability to recognize patterns and responsively modify their actions in a manner analogous to the way that humans learn from experience and practice.

The potential impacts of successful machine learning have seeped into popular culture. An article in *Wired* magazine announced, "The End of Code" (Tanz 2016) and

© The Author(s), under exclusive license to Springer Nature Switzerland AG 2023
J. Ganem, *Understanding the Impact of Machine Learning on Labor and Education*,
SpringerBriefs in Philosophy, https://doi.org/10.1007/978-3-031-31004-1_1

researchers at Google developed a computer program that beat a human grandmaster at the game of Go (Silver et al. 2017). The approach by the Go programmers was completely different than the one taken nearly twenty years earlier by the IBM team that built a computer that defeated the reigning world chess champion (Hsu 2002). Go is too complex a game to be approached through brute-force calculation and human experts have difficulty articulating their reasons for making particular moves. Human players must intuit their strategies through past experiences with many patterns. The "AlphaGo" computer taught itself to play Go by running many simulations and learning the patterns on its own. This self-taught approach has since been repeated with other strategy games, including chess, as well as shogi and Atari (Schrittwieser et al. 2020). By self-learning these kinds of strategy games, the computers also came to exhibit planning behaviors. This is a significant advance in AI because the inability to plan is what made it difficult for computers to beat humans in these kinds of strategy games, despite the enormous advantage computers have over humans in computational power.

That machines can learn, and even plan, and not need to rely on human programmers for all their actions promises to be more economically disruptive than the introduction of machines that merely performed work associated with intelligence. In the past few decades, the economics of industries as diverse as architecture, aviation, law, publishing, writing, accounting, manufacturing, and many others have been disrupted by the automation of routine tasks that in the past required a highly trained human expert. However human intervention in these activities is still required to deal with novel and/or unanticipated situations that cannot be predicted and pre-programmed.

The difference is that while machines and humans exhibit behaviors that are said to require "intelligence," humans have always had the ability to go a step further. Humans can change and adapt their thinking in response to external feedback as time progresses. This adaptive intelligence—the ability to learn—is fundamentally different than merely knowing. Because of the ability to learn, a human expert's "mental representations" (Ericsson and Pool 2016) in a specific field of expertise differ greatly from those of a human novice in the same field. In an analogous manner, a machine programmed with a learning algorithm will over time, modify its processing of inputs such that it produces different outputs as exposure time to the task increases.

The difference between AI and machine learning (a sub-branch of AI), and its profound social and economic implications, already has an extensive literature—see for example the historical survey *The Quest for Artificial Intelligence: A History of Ideas and Achievements* (Nilsson 2010). But new approaches to machine learning are prompting bold predictions for the future. *The Master Algorithm: How the Quest for the Ultimate Learning Machine Will Remake Our World*, surveys different approaches to machine learning, and argues that there must exist a single, yet to be discovered, learning algorithm that is universal (Domingos 2015). Domingos believes that *all* knowledge can be derived from data by a single universal learning algorithm and asserts that its discovery will have profound implications for business, science, and society.

Whether or not a universal learning algorithm exists and will be eventually discovered is outside the scope of this book. This book will examine potential disruptions in the labor market caused by machine learning and the implications for future labor and education policies. The arguments presented here apply to any kind of machine learning algorithm—whether it is an existing or yet-to-be discovered algorithm. It is assumed that advances in machine learning will continue and have profound impacts on the economy of the future. This book describes a modification of Turing's "Imitation Game" (Turing 1950), which is a proposed definition of machine thinking, as a way to frame learning, which it will be argued is fundamentally different than thinking because of the element of time. Thinking is a mental activity that occurs in the moment, while learning, which of course requires thinking, is an ongoing change in thought processes over time.

In Turing's original paper, his "Imitation Game," which has come to be known as the "Turing Test," provided an operational definition of machine thinking in terms of observed interactions with a human. Much has been written about whether operationalizing the observable outcomes that result from thinking is the same as actually thinking (Searle 1980). In addition, some philosophers argue that Turing's test is not an operational definition (Moor 2001). But those issues are outside the scope of this book. As is readily seen, a machine capability only needs to be operationalized for it to be economically disruptive.

This book is organized as follows:

Chapter 2. Labor Markets: Comparative Learning Advantages—argues that comparative labor advantages—as understood in standard economic texts (Samuelson and Nordhaus 2009)—arise from the disparate learning times for different occupations. It uses occupational and census data to show that the typical time required to learn a particular occupation (which will be defined as the "characteristic learning time") is predictive of earnings derived from that occupation. In general, the longer it takes to learn an occupation the greater the comparative labor advantage the practitioner accrues, which results in greater lifetime earnings relative to occupations that require less time to learn. However, just as machines have undermined the labor market by automating what were once valuable skills and knowledge, artificial learning threatens to undermine the relationship between learning times and wages. Workers have always had to learn new jobs as automation and technological change rendered jobs obsolete. However, for humans learning times are biologically limited. In principle, an artificial learning algorithm free from biological constraints can learn a new job in significantly less time than a human. It becomes possible to imagine a new job being automated by artificial learning faster than any human could learn the job. This undermines labor market models that predict that job losses caused by automation will be offset by the creation of new jobs that the automation enables (Acemoglu and Restrepo 2019).

To understand the idea that workers could be preemptively displaced by machines from new jobs, this chapter will introduce the concept of a "comparative learning advantage." This concept is analogous to a "comparative labor advantage" which arises from already possessed knowledgebase/skillsets that are used to perform tasks. But a comparative learning advantage arises from the time required to learn new knowledgebases/skillsets. This chapter argues that on a microeconomic-level an individual's comparative learning advantage determines his or her wage rate and on a macroeconomic-level comparative learning advantages determine the division of labor between humans. In the future, comparative learning advantages between humans and machines will determine their division of labor.

Chapter 3. Learning to Work: The Two Dimensions of Job Performance— examines what workers learn in order to be compensated in the labor market. It references an existing "task model" of production in which a "job" is the performance of a bundled sequence of discrete tasks (Autor et al. 2003). However, the tasks associated with different occupations have evolved over the last several decades as computer technology entered the workplace. Prior analysis of job task data dating back to 1960 has shown a shift away from routine tasks to non-routine and inter-personal tasks. This shift in job tasks has coincided with a shift in employer preference for college-educated workers who have comparative labor advantages over the non-college educated for nonroutine and interpersonal tasks.

This chapter sorts job tasks into two broad categories—those requiring *expertise* and those requiring *interpersonal* skills. Tasks that require expertise have stable endpoints, which makes these tasks inherently repetitive and subject to automation. Tasks that are interpersonal are highly context-dependent and lack stable endpoints, which makes these tasks inherently non-routine. Both expertise and interpersonal knowledgebase/skillsets are acquired through learning, which means that they take time and agency to obtain.

When labor is bought and sold in the marketplace, the participants in the transaction value knowing/performing, regardless of whether the knowledgebase/skillset the task requires is in the expertise or interpersonal category. How long it took a practitioner to learn his or her knowledgebase/skillset is not relevant to the transaction. Therefore, it becomes necessary to distinguish knowing/performing from learning. The difference is the element of time. Knowledge/performance levels for tasks requiring expertise can be ranked using time-independent distribution functions that are used to compute percentiles. By assigning percentiles, the knowledge/performance capability for an individual can be ranked relative to the entire population of practitioners. However, an individual through the intentional act of learning can over time move within a time-independent distribution. This movement is described by a time-dependent function called a "learning curve." Characteristic learning times can be derived from learning curves, but they cannot be derived from the time-independent distribution functions used to rank different levels of knowledge/performance. Learning is independent from knowing. As a result, learning cannot be ranked in the same way as knowledge/performance levels, but its speed and effectiveness can be assessed from the time-dependence of learning curves.

However, the process of learning and its assessment are different for expertise than for the interpersonal. Therefore, learning must be independently assessed along two dimensions—the expertise dimension and the interpersonal dimension. Traditionally, education focusses on teaching and assessing various areas of expertise. To date, machine learning has also primarily focused on acquiring expertise in various areas. Learning along the interpersonal dimension is much more difficult to teach and assess for people and machines. However, as interpersonal tasks become of greater value economically, this dimension cannot be ignored.

Chapter 4. The Judgment Game: The Turing Test as a General Research Framework—revisits Turing's three-participant "Imitation Game," that he argued could serve as a standard for establishing if machines can think, and formalizes the probabilistic and temporal nature of the game that Turing implied in his description but did not rigorously define (Turing 1950). This chapter will argue that Turing-like games can be used as general tests of distinguishability between levels of knowledge/performance in learned subject areas along the expertise and interpersonal dimensions.

It proposes a modification of the Imitation Game called the "Judgment Game," which also has three participants—an interrogator (Judge), who must distinguish between two players A & B. The players, despite inherent differences, attempt to be indistinguishable to the Judge. However, in the Judgment Game, the focus is on the success of the Judge, which can be scored and compared to other judges. Because the Judge's "job" is mostly interpersonal, such a game provides a method to assess knowledgebases/skillsets, along the interpersonal dimension, which has been problematic in the education of humans and machines. In fact, a machine in the role of the Judge in a Judgement Game, might be a higher standard for machine cognition than a machine in the role of Player A, as in Turing's original Imitation Game.

But it is argued that the Judgment Game, like the Imitation Game, evaluates performance; it does not evaluate learning. Therefore, these tests provide a restrictive definition of "thinking" because only thought processes that involve possession of prior, existing knowledgebases/skillsets are demonstrated. The thought processes involved with adaptive intelligence—learning—must have already taken place before these games can be played. In addition, it is possible for learning and forgetting to take place in the future, which means that these tests do not provide a stable standard to define "thinking."

Chapter 5. The Learning Game: A Time-Dependent Turing Test—presents a time-dependent version of the Judgment Game called the "Learning Game"— that permits two of the three participants to learn. The Judge is permitted to learn along the expertise and interpersonal dimensions, and Player A is permitted to learn along the expertise dimension, while the knowledgebases/skillsets of Player B are kept fixed to establish a comparison standard. In this game, the Judge's ability to distinguish between Players A & B becomes a complicated function of time, which means that different Learning Games can be constructed to simulate different learning environments. To illustrate, three special cases are considered: (1) An environment much like a typical education setting, in which the "Judge" is a teacher whose job is teach Player A (student) to become indistinguishable from Player B (graduate). (2) An environment in which the "Judge" is a manager whose job is to direct employees with much greater expertise. (3) An environment in which the "Judge" is a manager whose job is to direct employees with much less expertise. The time dependence of the Judge's score in these games provides feedback on the learning processes taking place. Therefore, Learning Games can be used as a training method for either Player A or the Judge, especially for acquiring knowledgebases/skills along the interpersonal dimension. It is also possible to substitute a machine for either Player A or the Judge and use a Learning Game to train the machine and assess its performance.

Chapter 6. Implications: Recommendations for Future Education and Labor Policies—The arguments presented in this book have significant implications for the future of work and education. A job in which there is a lack of time-dependence for the Judge's score can be modeled as Judgement Game, while the existence of a time-dependence for the Judge's score means that the job must be modeled as a Learning Game. Humans have a comparative learning advantage in Learning Games, while learning-enabled machines have a comparative learning advantage at Judgment Games. Therefore, humans should focus on jobs that can be modeled as Learning Games. Current education practices need revision because they focus on teaching knowledgebases/skillsets used in jobs that resemble Judgment Games. This contributes to the current mismatch in the labor market—in which many people are looking for work while, at the same time, many jobs are going unfilled. There are many other factors contributing to this labor market mismatch. In addition to education deficiencies, demographic shifts, disruptions from climate change, and barriers to immigration resulting from political dysfunction, all contribute to inefficiencies in the labor market. However, this book is about learning. It will focus on the education factor and make recommendations for changes to education practices.

Machine learning will fundamentally alter the relationship between humans and machines. Rather than productivity enhancement tools, machines will become, in a sense, co-workers and a division of labor will need to be negotiated. This book argues that comparative learning advantages between humans and learning-enabled machines will determine the division of labor. As a result, the future economy will be one in which human capital is valued for time-dependent-learning rather than static expertise that schools have traditionally instilled.

References

Acemoglu D, Restrepo P (2019) Automation and new tasks: how technology displaces and reinstates labor. J Econ Perspect 33(2):3–30. https://doi.org/10.1257/jep.33.2.3

Autor DH, Levy F, Murnane RJ (2003) The skill content of recent technological change: an empirical exploration. Q J Econ 118 (4):1279–1333. https://EconPapers.repec.org/RePEc:oup:qjecon:v: 118:y:2003:i:4:p:1279-1333

Domingos P (2015) The master algorithm: how the quest for the ultimate learning machine will remake our world. Basic Books

Ericsson KA, Pool R (2016) Peak: secrets from the new science of expertise. Houghton Mifflin Harcourt

Hsu FH (2002) Behind deep blue: building the computer that defeated the world chess champion. Princeton University Press

Moor JH (2001) The status and future of the Turing test. Mind Mach 11:77–93. https://doi.org/10. 1023/A:1011218925467

Nilsson NJ (2010) The quest for artificial intelligence. Cambridge University Press

Samuelson PA, Nordhaus WD (2009) Economics, 19th edn. McGraw-Hill Education

Schrittwieser J, Antonoglou I, Hubert T, Simonyan K, Sifre L, Schmitt S, Guez A, Lockhart E, Hassabis D, Graepel T, Lillicrap T, Silver S (2020) Mastering Atari, Go, chess and shogi by planning with a learned model. Nature 588:604–609. https://doi.org/10.1038/s41586-020-030 51-4

Searle J (1980) Minds, brains, and programs. Behav Brain Sci 3(3):417–424. https://doi.org/10.1017/S0140525X00005756

Silver D, Schrittwieser J, Simonyan K, Antonoglou I, Huang A, Guez A, Hubert T, Baker L, Lai M, Bolton A, Chen Y, Lillicrap T, Hui F, Sifre L, van den Driessche G, Graepel T, Hassabis D (2017) Mastering the game of Go without human knowledge. Nature 550:354–359. https://doi.org/10.1038/nature24270

Tanz J (2016) Soon we won't program computers. We'll train them like dogs. Wired, 17 May 2016. https://www.wired.com/2016/05/the-end-of-code/

Turing AM (1950) Computer machinery and intelligence. Mind LIX:433–460. https://doi.org/10.1093/mind/LIX.236.433

References

...

Chapter 2
Labor Markets: Comparative Learning Advantages

Abstract Comparative labor advantages arise from disparate learning times for different occupations. Occupational and census data are used to show that the typical time required to learn a particular occupation (defined as the "characteristic learning time") is predictive of earnings derived from that occupation. The longer it takes to learn an occupation the greater the lifetime earnings relative to occupations that require less time to learn—a result that is consistent with Human Capital Theory, which is a widely used framework in economics. However, machine learning threatens to undermine the relationship between learning times and wages because for humans, learning times are biologically limited while for machines, a new job could be learned in significantly less time than a human. This undermines labor market models that predict that job losses caused by automation will be offset by the creation of new jobs that the automation enables because workers could be preemptively displaced by machines. To understand this effect the concept of a "comparative learning advantage" is introduced. This concept is analogous to a comparative labor advantage which arises from already possessed knowledgebase/skillsets used to perform tasks. But a comparative learning advantage arises from the time required to learn new knowledgebases/skillsets.

Keywords Comparative labor advantage · Comparative learning advantage · Worker displacement by job automation · Human Capital Theory · Effect of workplace automation on wages · Dependence of lifetime earnings on education

2.1 Opportunity Costs for Labor Are Temporal

> I have learned that the swiftest traveler is he that goes afoot. I say to my friend, suppose we try who will get there first. The distance is thirty miles, the fare is ninety cents. That is almost a day's wages. I remember when wages were sixty cents a day for laborers on this very road. Well, I start now on foot, and get there before night; I have travelled at that rate by the week together. You will in the meanwhile have earned your fare, and arrive there sometime tomorrow or possibly this evening, if you are lucky enough to get a job in season. Instead of going to Fitchburg, you will be working here the greater part of the day. (Thoreau 1854, p. 68)

This passage in Thoreau's famous book, *Walden*, on his experiment living by himself from 1845 to 1847 in the woods near Walden Pond outside Concord, Massachusetts, shows that he understood the economic concept of "opportunity costs," and that in the case of human labor, opportunity costs are measured in time. Laboring a day or more to earn the train fare to Fitchburg would require more time than walking to Fitchburg. Forgoing a day or more of wages to walk to Fitchburg—an action that requires roughly the same physical exertion as being a day laborer—instead of paying the train fare is a profitable trade. A day spent walking is an "opportunity cost," because it costs the opportunity to earn a day's wages, but the gain—the ninety cents saved by not paying the train fare—is greater than this cost. When the total time to get to Fitchburg is summed—working plus riding compared to walking—the walker wins the race.

Of course, the outcome of the race to Fitchburg in the above passage would be completely different if Thoreau could earn in one day, wages that are greater in comparison to the train fare. Suppose Thoreau's daily wage was $2 per day. Now the train is faster because he would only need to work half a day to pay the fare, but it would still take a full day to walk. Under these circumstances, an exchange of labor for the train ticket would be profitable. The worker-train rider would beat the walker.

If Thoreau lived in modern times and wanted to go to Los Angeles, he would find working to buy a plane ticket and then flying, much faster than walking. Even if he were to earn the federal minimum wage of $7.25 per hour it would only take him about a week's wages (40 hours) to purchase an airplane ticket for a 5.5-hour flight from Boston to Los Angeles. However, according to *Google Maps*, walking this 2963-mile distance requires 974 hours. Thoreau's assertion that "the swiftest traveler goes afoot" completely breaks down because of the change in opportunity costs brought about by the technological disruption of modern air travel.

2.2 Comparative Advantages for Labor

Standard economic texts show that because of opportunity costs it is not necessary for there to be an absolute advantage in labor, capital, or resources for trade between two or more parties to be mutually beneficial (Samuelson and Nordhaus 2009). All that needs to exist is a "comparative advantage." For example, suppose Alice is both a much better heart surgeon *and* a much better plumber than Bob. But if heart surgeons are paid much more than plumbers it is still better for Alice to hire Bob to do her plumbing. If Alice is paid $300 per hour to do heart surgery, and Bob is paid $100 per hour to do plumbing, Alice can do an hour of heart surgery, pay Bob $200 for two hours of plumbing and pocket the $100 difference. Even if she could have performed Bob's two-hour plumbing job in one hour because of her superior plumbing skills, she is still ahead on this transaction because she would have given up $300 to save herself $200. The trade works because Bob can't do heart surgery no matter how much he is paid. The opportunity cost to Alice of not doing heart surgery gives Bob a comparative advantage in plumbing. Alice may be a better plumber than Bob, but the lost opportunity to do heart surgery is a cost to her that benefits Bob.

Comparative advantages resulting from disparate opportunity costs are inherently fractal, in that they can appear at all scales—for individuals, businesses, industries, and even entire countries. It is not necessary for a particular entity—no matter the size—to have an absolute advantage for trade to be mutually beneficial—only a comparative advantage arising from differences in opportunity costs.

Other kinds of economic resources—not just human capital—are subject to disparate opportunity costs. Consider the allocation of financial capital. An investor choosing to purchase stock in corporation ABC is foregoing the opportunity to invest in XYZ. If XYZ gains 50% over the next four years while ABC only gains 10%, the investor's choice—even though profitable—still had an opportunity cost in that it lost the opportunity for the higher rate of return. Infrastructure spending also involves opportunity costs. A municipality's decision to upgrade its airport as opposed to its public transportation network will impact the composition of its future workforce and businesses. Its future economy may depend heavily on airline-related businesses, while a neighboring municipality that chose to upgrade its public transportation network might find that it has a comparative advantage in attracting manufacturing businesses.

In all these examples, it is not just the scarcity of the resource that affects the decision, but as Thoreau well understood, the element of time. Opportunity costs are not just financial—they are also temporal. The rate of return on an investment is measured in time. Present infrastructure spending will determine future economic opportunities. For human labor, time is an especially important element of opportunity costs because it is fixed. An individual can grow access to capital and material resources with no relevant upper bound[1] in place, but there is never more than 24 hours in a day, and a human lifetime will not—as of this writing—significantly exceed 100 years. In many ways, time is the scarcest of all economic resources because for a given individual, unlike available capital, it cannot be increased.

2.3 Automation Disrupts Opportunity Costs for Labor

Because time is so strictly constrained for human activity, comparative advantages for exchange of labor can be calculated for the example of Alice and Bob—or any other group of wage earners—just by knowing wage-rates. Alice cannot add more hours to her day in the same way that she could potentially own stocks in both ABC and XYZ corporations, which makes the calculation of opportunity costs for financial capital more complicated.

---

[1] Obviously the capital an individual can acquire is bounded by the total capital available. However even the wealthiest individual on the planet still possesses a relatively small fraction of the total wealth available. Therefore, wealth disparities can be enormous—many orders of magnitude separate the wealthiest from the poor, but human adult lifetimes can never differ by even one order of magnitude.

As Martin Ford explains:

> Comparative advantage works because of opportunity cost: if a person chooses to do one thing, she must necessarily give up the opportunity to do something else. Time and space are limited: she can't be in two places doing two things at once. (Ford 2015)

However, automation—the creation of a robotic Alice—disrupts the usual economic exchanges of human labor. A robot version of Alice, to use Martin Ford's terminology, is effectively a "clone." If Alice can be cloned, she could be doing heart surgery and plumbing during the same hour. A robot with Alice's superior surgical and plumbing skills that can be replicated, mass-produced, and deployed in operating rooms and construction sites is the equivalent of cloning Alice. The question now becomes how much does it cost to clone Alice—that is create and operate a robot with her skills? If the time-average cost to manufacture, maintain, and operate that robot is $50 per hour for the plumbing feature there will be downward pressure on Bob's pay. To compete with the robot, he will not be able to charge more than $25 per hour (remember the robot has Alice's 2 to 1 time-advantage in plumbing ability). If the time-average cost is $500 per hour for the robot to have the heart surgery feature, Alice will still be able to charge her $300 rate, at least for the time being. The history of automation is that over time costs fall. Jobs that initially appear safe from automation, are often not safe in the future.

Therefore, if Alice can be cloned at a cost that is small compared to employing Bob, the concept of opportunity cost for labor is no longer relevant, and Bob is out of a job. The comparative advantage principle for labor no longer applies. This unravels the entire labor market because the modern economy is built on increasing divisions of labor that incentivize workers to become more and more specialized in order to maximize their comparative advantages. But, as Ford points out, paradoxically the more specialized work becomes the more susceptible it is to automation. A robotic Alice doesn't need her general intelligence, it just needs her specialized, narrowly defined skillsets to put Bob and eventually Alice out of their jobs.

Technology is maturing to the point that *any repetitive task* can be automated—including those that in past required high-levels of expertise. Machines have already replaced workers in many low-skilled jobs in manufacturing and retail. But now data-fed algorithms generate natural language sports stories and stock market reports for publication. There are robotic investment and legal advisors. It is expected that as machine vision and image processing improve, autonomous vehicles will progress from experimental demonstrations to widespread use. Radiologists could be replaced by algorithms that interpret x-rays more accurately. And yes, even surgeons like Alice might be replaced by robots.

In more general terms, automation enables capital to compete with and displace labor, an effect that depresses wages. As a result, there has been no upward pressure on the minimum wage for decades. The real value, adjusted for inflation, of the federal minimum wage peaked in 1968 and has been in decline ever since (Center for Poverty & Inequality Research 2018). This more than five-decade decline coincided with the widespread dissemination of digital technology into workplaces and the automation of many low-wage jobs that this technology enables. This trend is consistent with economic models that predict that increasing the minimum wage results in a significant decrease in the share of automatable employment held by low-skill workers (Lordan and Neumark 2018) and high-skill workers (Acemoglu and Restrepo 2018a).

While automation creates a powerful displacement effect for workers, economists argue that the net effect on the labor market is offset by a productivity effect—that is automation results in the creation of new tasks in which labor has a comparative advantage. A framework for modeling these effects (Acemoglu and Restrepo 2018b), shows that this productivity effect can offset the displacement effect. They state: "… the biggest shortcoming of the alarmist and the optimist views is their failure to recognize that the future of labor depends on the balance between automation and the creation of new tasks." However, they go on to recognize that there is an inherent inefficiency in the labor market for the new tasks because there is a mismatch between the knowledge/skills that the new tasks require with the actual skills of the workforce. As a result, the theoretical balance between the automation of old tasks and the creation of new ones is hindered by an adjustment process—that is the time and effort it takes workers to learn the new tasks. They write:

> … the adjustment process is likely to be slower and more painful than this account of balance between automation and new tasks at first suggests. This is because the reallocation of labor from its existing jobs and tasks to new ones is a slow process, in part owing to time-consuming search and other labor market imperfections. But even more ominously, new tasks require new skills, and especially when the education sector does not keep up with the demand for new skills, a mismatch between skills and technologies is bound to complicate the adjustment process. (Acemoglu and Restrepo 2018b)

This means that bottleneck is not the rate at which new tasks are created, *but the rate at which the workforce can learn them.* In other words, if Bob is displaced by a plumbing robot, it will take him time and cost him money to learn a new trade. Humans need time to learn because, unlike an algorithm, their learning is biologically limited and cannot be sped up. In addition, learning is rarely obtained for free.

2.4 Characteristic Learning Times and the Effect on Wages

Suppose Bob loses his comparative advantage over Alice in plumbing to a robot. An option for him is to spend time and money to learn another trade. If he chooses that strategy, it is worth it for him to consider why he had a comparative advantage over Alice in plumbing. That reason is intimately connected with time—there are two very different time scales associated with learning plumbing as opposed to heart surgery. With a high school diploma and a two-year apprenticeship, Bob was installing plumbing fixtures at age 20, just 4 years after becoming work-eligible at age 16. Alice's occupation required four years of college, four years of medical school and a two-year residency. She didn't graft bypasses on her own until age 28. She is paid more because it took her more time to acquire surgery skills and as a result, she will always be better off financially by outsourcing her plumbing needs to Bob.

Similarly, neither Bob nor Alice will ever find it worth their time to compete with Cathy the cashier for her job. Cathy became a cashier as soon as she was work-eligible at age 16 with no formal training at all. No matter how much better Bob or Alice might be than Cathy at being a cashier, her job is safe from them—although it is not safe from automation. Even though it is the lowest paid job which provides the least incentive to automate, it is also the easiest to automate because it is the most repetitive. In addition, there are many more cashiers than surgeons, so cloning Cathy would benefit from economies of scale.

Every occupation requires time for a human to learn. However, learning times vary greatly depending on the extent and complexity of the knowledgebases/skillsets the occupation demands. The average time it takes for a person to acquire the skills and credentials necessary for a particular occupation will be defined as the "characteristic learning time"—T_c. Heart surgery, for example, has a much longer T_c than plumbing and because this surgical skillset takes longer to acquire it pays more. That Alice is paid more than Bob because the T_c for heart surgeons is longer than the T_c for plumbers is a general pattern. While it is difficult to determine the characteristic learning time for every occupation and relate it to its wage rate, one proxy for this relationship is to plot the years of formal education required for entry to an occupation versus its median annual wage. In the following analysis "years of required formal education for entry-level" serves as a proxy for "characteristic learning time"—T_c.

The U. S. Department of Labor's Bureau of Labor Statistics maintains an online *Occupational Outlook Handbook*, that provides descriptions, entry-level education requirements, and median annual pay for about 300 different occupations (Bureau of Labor Statistics 2019). Figure 2.1 shows a scatter plot of 2018 median annual pay in dollars versus years of entry-level education required (T_c) for 282 occupations published in 2019.[2] The few occupations that had no explicit entry-level education requirements and/or a median annual pay published are excluded from Fig. 2.1 plot. Years of entry-level education required, used as a proxy for T_c, are coded as in Table 2.1.

Figure 2.1 plot in shows a strong correlation between T_c for an occupation and its median annual wage. A linear regression gives the following relationship for W (the median annual wage in 2018 dollars) and T_c (the characteristic learning time in years from Table 2.1):

$$W = \$25,000 + \$8600\,T_c \tag{2.1}$$

Despite the scatter in Fig. 2.1 plot, a statistical "F-test" performed on the data showed that the linear relationship (Eq. 2.1) is highly significant and unlikely to

---

[2] The *Occupational Outlook Handbook* maintained by the Bureau of Labor Statistics is an online interactive publication that is updated each year to reflect the prior year's median pay and ongoing changes in job outlook and demand. The data and analysis presented in Tables 2.1 and 2.2 and Fig. 2.1, uses 2018 data published and accessed in 2019.

Fig. 2.1 Median annual wage (2018 dollars) versus T_c (the characteristic learning time in years from Table 2.1). A linear fit to this data results in Eq. (2.1)

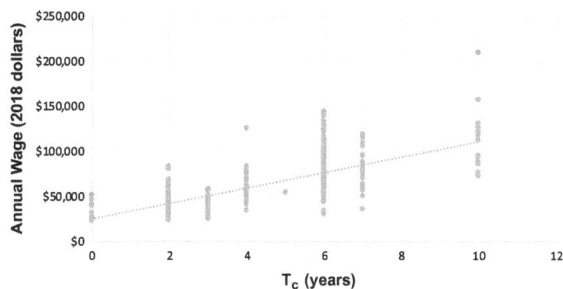

Table 2.1 Coding of T_c for different entry-level education requirements

Entry-level education requirement	T_c (years)
No formal education	0
High school diploma	2
Some college or a non-degree certificate	3
Associate degree	4
Bachelor's degree	6
Master's degree	7
Juris doctorate	9
Doctorate or professional degree	10

The T_c time codes are based on the typical coincidence of age-16 work eligibility with progress in age-appropriate formal education—e.g. compulsory school attendance ends at age 16, high school graduation at age 18, a two-year associate degree at age 20, and so on

occur by chance. While this is a simple model and it uses formal education as a proxy for T_c, Eq. (2.1) can be used to make accurate predictions of expected lifetime earnings based on educational attainment as determined from U. S. Census Bureau survey data. Consider Table 2.2, which in the first three columns respectively show educational attainment, associated T_c, and annual wage rate for that T_c given by Eq. (2.1). Expected lifetime earnings (shown in the fourth column), assuming a forty-year career and an annual age wage rate given by Eq. (2.1), are calculated by simply multiplying 40 years times the annual wage rate.

For comparison, calculations of "synthetic work-life earnings by educational attainment" taken from a 2012 U. S. Census Bureau report are shown in the adjacent fifth column. According to the report (Julian 2012): "Synthetic work-life earnings represent expected earnings over a 40-year period for the population aged 25–64 who maintain full-time, year-round employment the entire time. Calculations are based on median annual earnings from a single point in time for eight 5-year age groups and multiplied by five." The data for this calculation comes from *The American Community Survey*, administered in 2011 by the U. S. Census Bureau.

Table 2.2 Expected lifetime earnings by educational attainment

Educational attainment	T_c (Table 2.1)	W = Annual wage rate in 2018 dollars given by Eq. (2.1)	40-year earnings in 2018 dollars = 40 * W	40-year earnings (2011 U. S. Census Data)	Ratio of 40-year earnings to a high school graduate given by Eq. (2.1)	Ratio of 40-year earnings to a high school graduate (2011 U. S. Census Data)
No Formal education	0	25,000	1,000,000	1,017,500[a]	0.59	0.74
High school diploma	2	42,200	1,688,000	1,371,000	1.00	1.00
Some college	3	50,800	2,032,000	1,632,000	1.20	1.19
Associate degree	4	59,400	2,376,000	1,813,000	1.41	1.32
Bachelor's degree	6	76,600	3,064,000	2,422,000	1.82	1.77
Master's degree	7	85,200	3,408,000	2,834,000	2.02	2.07
Doctorate or professional degree	10	111,000	4,440,000	3,842,000[b]	2.63	2.80

[a] Average of the earnings for the "None to 8th grade" and "9th to 12th grade" categories = (936,000 + 1,099,000)/2
[b] Average of the earnings for the "Professional degree" and "Doctorate degree" categories = (4,159,000 + 3,525,000)/2

Of course, the absolute dollar amounts for expected 40-year earnings are much larger for 2018 wages than for those based on 2011 wages and this is seen in Table 2.2. Therefore, to express the effect of educational attainment (the proxy for T_c) on lifetime earnings independent of the year wages were sampled, ratios are calculated using the respective high school earnings figures as the denominator. These ratios are shown in the two leftmost columns. These ratios are very nearly the same in the two columns. This shows that Eq. (2.1) is highly predictive of the effect of educational attainment on lifetime earnings as reported by the U. S. Census Bureau survey.

The U. S. Census Bureau report also acknowledges what is effectively the reason for the large scatter in Fig. 2.1 from which Eq. (2.1) is inferred:

> Not everyone working in the same occupational category with the same level of education earns the same amount. For example, workers who majored in engineering make an estimated $3.3 million in their work life, while arts majors make $1.9 million even though it took them just as long to learn their trades. (Julian 2012)

The market does indeed value some skills more than others even when T_c's are roughly comparable. It might take just as much time to learn art as engineering, but the market values the engineer's skillset more.

2.5 Human Capital Theory

This analysis in Sect. 2.4 is consistent with Human Capital Theory (HCT), which is a framework in economics that originated in the mid-twentieth century and profoundly influenced modern education policy. HCT sought to address an economic paradox—that abilities are normally distributed within a population, but personal incomes are highly skewed (Mincer 1958). This implies that a normally distributed personal attribute—such as IQ—does not account for personal income. This led to the idea that the knowledgebases/skillsets that people intentionally acquire at a cost, are forms of capital. The cost of education becomes framed as a deliberate investment. As explained by Theodore Schultz:

> Much of what we call consumption constitutes investment in human capital. Direct expenditures on education, health, and internal migration to take advantage of better job opportunities are clear examples. Earnings foregone by mature students attending school and by workers acquiring on-the-job training are equally clear examples. (Schultz 1961)

Over the next few decades, the idea that education is an investment in "human capital" evolved into a framework (HCT) used to study the economics of education. Jacob Mincer, a founder of the field, went on to construct a sophisticated mathematical model—the Mincer Earnings Function (Mincer 1974). His model explained the skewed distribution of lifetime earnings, taken from 1950s era census data, in terms of work experience and life-cycle investment in education. The exact form of the Mincer Earnings Function depends on the assumptions made about the time-value of earnings and educational investments. But, with a reasonable set of assumptions it predicts that much of the observed income inequalities arise from differences in education. My crude model above is less sophisticated mathematically, but the qualitative predictions are similar, and it is consistent with the much more recent census data.

Historical articles "Human Capital" (Eid and Showalter 2010) and "The Introduction of Human Capital Theory into Education Policy in the United States (Holden and Biddle 2017), show how HCT transformed the public discourse around educational policy. Rather than a public good, education became framed as an investment that, if done wisely, would result in greater economic productivity. Essentially, an educated workforce becomes part of the nation's infrastructure—analogous to roads bridges—or at least treated like infrastructure in the economic models used. It is interesting to note, in the quote above from one of HCT's foundational papers, Schultz mentions "health" as an example of investment in human capital. Yet, HCT did not transform healthcare policy in the United States in the way that it did education policy.

Despite its widespread usage, HCT is not without critics. For an organized summary, with references, of the many criticisms see "Human Capital Theory: A Holistic Criticism" (Tan 2014). Some of these criticisms question the methodology of HCT, some question its neoclassical foundations in Rational Choice Theory, and some question HCT on moral grounds. The concept of people as infrastructure and education solely as an economic investment is dehumanizing. In addition, by focusing on individuals, HCT does not account for structural inequalities in society that produce unequal educational experiences and outcomes. Without that social context, it is easy to misuse HCT to support "blame-the-victim" style arguments.

The purpose in this book is not to critique HCT or expand on it. Nor is it to analyze the moral failings at a societal level that result in unequal education outcomes and earnings. The analysis in the preceding section, which is consistent with HCT, documents as a fact that people with education, on average, earn more than people without an education. It is also a fact that additional time devoted to education tends to correlate with additional earnings. This strong correlation is likely both directly causal—employers pay more for skills obtained through education—*and* the result of structural inequalities caused more by moral failures within society (e.g. underfunded segregated schools) than by moral failures of the individual (e.g. a personal choice not to work or to become educated). Rather, this book will argue that machine learning has the potential to break the correlation between education and earnings, which could make HCT irrelevant, at least in regard to the current ways that marketable skills are understood.

2.6 Individual Learning Times and Comparative Learning Advantage

The preceding analysis shows, that as a rule, low-wage earners obtain their comparative advantages over high-wage earners because the latter must spend much more time learning their occupations. This effect extends well beyond the years of formal education because of the widespread practice of paying more to experienced workers and basing future pay on prior pay. The market confers higher value on labor provided by individuals with knowledge, skills, and experience that are time-consuming to acquire. The more time an individual spends learning an occupation, the higher his or her wages are relative to other people in the labor pool, which results in strong economic incentives for that individual to outsource work not related to his or her occupation—even if that individual could perform that outsourced work just as well if not better.

However, and this is a critical point, T_c for an occupation is not necessarily the same for all individuals. Each person has an "individual learning time"—designated T_i—associated with each occupation. Not everyone capable of completing college can do so by age 22. Not everyone capable of learning to be a heart surgeon can complete a medical degree with four years of post-college study. Suppose Bob is just as capable of learning heart surgery as Alice, but it would take him much longer. Imagine that he can acquire Alice's skill set but it would take him 5 years to complete college and 6 years to complete medical school. Alice might be faster at learning plumbing, but not that much faster—say she could complete a two-year plumbing apprenticeship in 18 months. Given these differences in learning aptitudes and abilities, Bob and Alice will never compete with each other for medical school admission, or for the same plumbing apprenticeships, even before any comparative labor advantage between them for plumbing and heart surgery develops.

Regarding plumbing, Bob has what will be termed a "comparative learning advantage." *A comparative labor advantage between Bob and Alice cannot exist until each has learned his and her respective occupation.* Alice and Bob, like all other wage earners, must choose career paths and that choice often depends on how his or her individual T_i compares to the T_c for the occupation chosen. This is because the act of learning an occupation is itself an opportunity cost. Time spent learning is time not spent earning money, and in addition to lost wages, education usually requires a financial expenditure. The longer it takes to learn—meaning the greater T_i—the greater the opportunity costs associated with education, which will be a subtraction from the individual's expected future earnings. Market wage rates, however, are determined by T_c, not T_i. A plumber who takes twice as long to learn the trade as the average plumber will not be able to charge more to recover the greater opportunity costs; neither will a heart surgeon that takes twice as long to finish medical school be able to charge more. As a result, an individual usually does not choose an occupation for which his or her T_i is significantly greater than the T_c for the occupation. It is the difference between the T_i's, that are unique to each individual and T_c, which is a single number associated with each occupation, that result in comparative learning advantages.

There are potentially many reasons that T_i's for a given occupation vary between individuals. People can differ in innate talent, interests, and dispositions. Prior life experiences can pre-dispose some individuals to learning a particular skill faster than others. There are also structural and demographic issues within society—there are populations of people who are never provided the opportunity to acquire the educational background necessary for learning some occupations in a reasonable timeframe. In a just and equitable society, with equal educational opportunities, the variation in T_i's should be reduced. However, because of intrinsic differences between individuals, a variation will always exist.

2.7 Relevant Times and Expected Earnings

Comparative labor advantages are intimately connected to the disparate learning times for different occupations and employment positions. T_c determines the market wage rate for an occupation according to Eq. (2.1), while the comparisons of T_i's to the T_c's determine the comparative learning advantages between individuals. However, the analysis of expected lifetime earnings by the U. S. Census Bureau summarized in Table 2.2 is based on two assumptions that were roughly valid in the past, but in the future, as more work is automated and AI becomes more adept at learning, will no longer be valid.

1. It assumed a forty-year career beginning at age 25—nine years after typical work eligibility and ending at retirement age 65. This neglects the opportunity costs incurred acquiring education and training that often occurs during the nine years from age 16 to 25—costs that cannot be neglected as education expenses have rapidly increased in recent decades and continue to do so.
2. It assumed the education and training acquired will be relevant for forty years. Costs incurred to retrain and repurpose as industries, technologies and occupations radically change or become obsolete are neglected. Given the predictions of labor market models (Acemoglu and Restrepo 2019) that automation leads to the creation of new tasks, machine learning will undoubtedly accelerate this process and render valuable skillsets irrelevant on time scales much less than 40 years.

Therefore, a third time scale must be defined—the "relevant time"—T_r. This is the time that a knowledgebase/skillset for a given occupation is relevant. If a plumbing robot displaces Bob early in his planned 40-year career, the lifetime earnings he expected when he learned plumbing will never materialize. In addition, if he decides to learn a new trade—electrician for example—the expected earnings from that trade will not materialize if an electrician robot displaces him in the near future. The appearance of such robots on short time scales might also disrupt the simple linear relationship between wages and T_c's expressed in Eq. (2.1), although it is still likely that future wages will depend in some way on T_c. Because opportunity costs for learning and expected earnings are unique to the person, an individual's total expected earnings E_i from an occupation with a given T_c is given by the relation:

$$E_i = W(T_c) * T_r - C_O(T_i) - C_E(T_i) \qquad (2.2)$$

In this expression $W(T_c)$ is the annual wage that can be approximated by Eq. (2.1), but in principle can be a more complicated function of T_c. The function C_O is the opportunity cost for learning (e.g. lost wages while a person is in school) and C_E the direct education cost (e.g. tuition, supplies, etc.). Both are functions of T_i (because learning times and costs vary between individuals) and are negative (subtractions from the expected earnings). The positive contribution to E_i is W multiplied by T_r (not 40 years as per the calculations for Table 2.2) because if a time arrives that the acquired knowledgebase/skillset is no longer relevant, no further earnings can be derived from that occupation even if retirement at age 65 is years away.

Figure 2.2 shows plots of Eq. (2.2) as a function of T_r for four different T_c's with the following idealized assumptions:

- $C_E = 0$ (free education)
- $W(T_c)$ is given by Eq. (2.1)
- $T_i = T_c$ (the individual learns the occupation in the characteristic time)
- $C_O(T_i)$ is lost wages (given by Eq. 2.1) with $T_c = T_i$.

Fig. 2.2 E_i (2018 dollars) as a function of T_r (years) for four different T_c's calculated using Eq. (2.1). The four curves shown are $T_c = 0$ (no formal education), $T_c = 2$ years (high school diploma), $T_c = 6$ years (bachelor's degree), and $T_c = 10$ years (doctorate/professional degree)

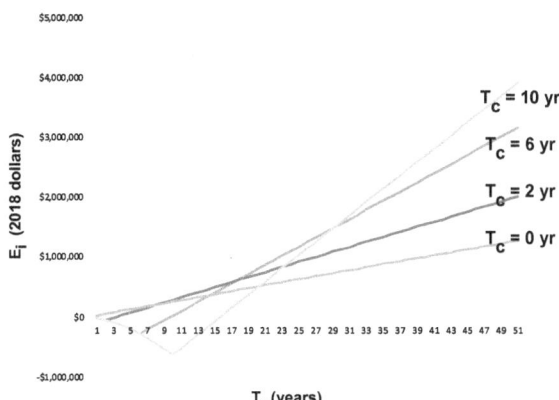

If time $= 0$ in Fig. 2.2 is age 16 and the knowledgebase/skillset acquired are relevant until full retirement at age 66 ($T_r = 50$), then the premium on total earnings expected from longer education times (T_c) shown in Table 2.2 will be realized. However, for T_r of approximately 15–20 years learning an occupation with greater T_c has no significant impact on total expected earnings. For $T_r = 10$ years or less, it is more beneficial financially to choose an occupation that requires no formal education ($T_c = 0$ has greater or equal expected earnings compared to the others).

In fact, unless T_r is approximately $2T_c$ no financial gains from education are available, which means that under that condition workers have no financial incentive to learn a new occupation if the learning requires any significant time or expense. In addition, disruption caused by automation and technological breakthroughs is inherently difficult to anticipate and predict, which makes knowing T_r for any specific occupation impossible. One trend that is clear, the T_r's for many learned knowledgebases/skillsets are becoming much less than 40 years. Worse, if machine learning creates an economy in which T_r for many occupations becomes less than T_c, workers have no possibility of using education to enhance future earnings. This all assumes an ideal world with free education. The reality—that education is expensive—only means that T_r must be even longer than $2T_c$ to realize any earnings benefit from education.

Economic models that predict that workers displaced by automation will adapt by switching to the new tasks created by the enhanced productivity run into a fundamental limitation of the human mind—it takes time to learn. In addition, unlike for machines, human learning times cannot be sped up. Human development unfolds according to a relatively rigid time schedule. Yes, there are wide variations in T_i's for different tasks. But T_i's in relation to T_c's do not vary by orders of magnitude. Alice might be a fast learner in medical school, but she will not be doing heart surgery after one month of training. But in principle, a robotic Alice could learn heart surgery in one month, or one day, or even less time.

2.8 Summary

In discussing the time-dependence of learning, three distinct time scales are distinguished.

1. Characteristic learning time (T_c)—the average time to learn the occupation and/or acquire the necessary knowledgebases/skillsets.
2. Individual learning time (T_i)—the time it takes a particular individual to learn the occupation and/or acquire the necessary knowledgebases/skillsets.
3. Relevant time (T_r)—the time into the future over which the knowledgebases/skillsets have economic value.

At the macroeconomic-level, comparative advantages in the labor market arise from characteristic learning times $(T_c$'s). Plumbers are paid less than surgeons because the characteristic learning time for plumbing is much less than the characteristic learning time for surgery. This is the pattern found in Fig. 2.1. The greater T_c, the greater the median annual wage, and workers in occupations with shorter T_c's who are paid less enjoy a comparative advantage over their higher earning counterparts.

At the microeconomic-level, comparative advantages arise from individual learning times $(T_i$'s). Alice might be able to learn plumbing twice as fast as Bob, but she might be 10 times faster at learning surgery. Bob found the pre-med college curriculum so overwhelming he never even considered it as a career option and thus never competed with Alice at all for medical school entrance. Bob does not have an absolute advantage for individual learning times, he has a comparative learning advantage.

Because of the opportunity costs incurred in learning, individuals will generally choose an occupation in which their T_i is roughly equal to the T_c for that occupation. The choice of an occupation in which an individual's T_i is much greater than the T_c for the occupation is generally not an economically viable choice because the opportunity costs associated with learning might not be recoverable. Maybe Bob could learn surgery if he spent an additional 5 years beyond the 10-year T_c for that occupation. But, that additional 5 years of learning costs him 5 years of working and the extra effort will not result in greater earnings when he completes the training. It is T_c that determines the market rates for labor (Fig. 2.1), not T_i. The time it takes a particular individual to learn an occupation will not change the market price for labor in that occupation.

Both individual and characteristic learning times are constrained by the relevant time (T_r) that a knowledgebase/skillset has economic value. At the most extreme, the relevant time cannot exceed an individual's lifetime. However, humans are now living in a period of history in which relevant times are often much less than human lifetimes—rendered much less by rapid increases in both automation and obsolescence. As a result, it has become increasingly necessary for individuals to acquire new skills and enter new occupations throughout the course of a lifetime. However, the need to change occupations does not change T_c, or the economic value of the labor associated with T_c because the rate at which humans learn cannot be significantly increased.

To summarize, an individual in the process of choosing an occupation can expect future employment, positive total earnings, and a comparative labor advantage provided that the following inequality holds.

$$T_i \leq T_c \leq 2\,T_r \tag{2.3}$$

This inequality provides two criteria to consider in the decision to learn a particular occupation:

1. $T_c \leq 2T_r$ (Employment condition)
2. $T_i \leq T_c$ (Comparative learning advantage condition).

Or in terms of Bob's predicament: (1) If the plumbing robot is causing him to consider doing two years of training to become an electrician, an electrician robot should be more than four years in the future, or Bob will not be able to recover his opportunity costs incurred by the training because the electrician robot will displace him before he can realize positive total earnings. (2) If it takes Bob longer than the two-year characteristic time to learn the electrician trade, he will increase his opportunity costs incurred by learning and thus increase his risk of not recovering them. In other words, he will not have a comparative learning advantage.

References

Acemoglu D, Restrepo P (2018a) Low-skill and high-skill automation. J Hum Cap 12(2):204–232. https://doi.org/10.1086/697242

Acemoglu D, Restrepo P (2018b) Artificial intelligence, automation and work. NBER Working Paper No. 24196. https://doi.org/10.3386/w24196

Acemoglu D, Restrepo P (2019) Automation and new tasks: how technology displaces and reinstates labor. J Econ Perspect 33(2):3–30. https://doi.org/10.1257/jep.33.2.3

Bureau of Labor Statistics (2019) Occupational outlook handbook. U. S. Department of Labor, Retrieved 11 Aug 2019, from https://www.bls.gov/ooh/home.htm

Center for Poverty & Inequality Research (2018) What are the annual earnings for a full-time minimum wage worker? University of California Davis. Retrieved 18 Aug 2021, from https://poverty.ucdavis.edu/faq/what-are-annual-earnings-full-time-minimum-wage-worker

Eide ER, Showalter MH (2010) Human capital. In: Peterson P, Baker E, McGaw B (eds) International encyclopedia of education, 3rd edn. Elsevier, pp 282–287. https://doi.org/10.1016/B978-0-08-044894-7.01213-6

Ford M (2015) The rise of the robots: technology and the threat of a jobless future. Basic Books

Holden L, Biddle J (2017) The introduction of human capital theory into education policy in the United States. Hist Polit Econ 49(4):537–534. https://doi.org/10.1215/00182702-4296305

Julian T (2012) Work-life earnings by field of degree and occupation for people with a bachelor's degree: 2011. American Community Survey Briefs ACSBR/11-04. U. S. Department of Commerce Economics and Statistics Administration, U. S. Census Bureau. https://www.census.gov/library/publications/2012/acs/acsbr11-04.html

Lordan G, Neumark D (2018) People versus machines: the impact of minimum wages on automatable jobs. Labour Econ 52:40–53. https://doi.org/10.1016/j.labeco.2018.03.006

Mincer J (1958) Investment in human capital and personal income distribution. J Polit Econ 66(4):281–302. https://www.jstor.org/stable/1827422

Mincer J (1974) The human capital earnings function. In: Schooling, experience, and earnings. National Bureau of Economics Research, pp 83–96. http://www.nber.org/books/minc74-1

Samuelson PA, Nordhaus WD (2009) Economics, 19th edn. McGraw-Hill Education

Schultz TW (1961) Investment in human capital. Am Econ Rev 51(1):1–17. https://www.jstor.org/stable/1818907

Tan E (2014) Human capital theory: a holistic criticism. Rev Educ Res 84(3):411–445. https://www.jstor.org/stable/24434243

Thoreau HD (1854) Walden, Apollo edn. Thomas Y. Cromwell, p 68

Chapter 3
Learning to Work: The Two Dimensions of Job Performance

Abstract Job tasks—performed by workers to be compensated in the labor market—are sorted into two broad categories—those requiring *expertise* and those requiring *interpersonal* skills. Tasks that require expertise have stable endpoints, which makes these tasks inherently repetitive and subject to automation. Tasks that are interpersonal are highly context-dependent and lack stable endpoints, which makes these tasks inherently non-routine. Both expertise and interpersonal knowledge-base/skillsets are acquired through learning, which means that they take time and agency to obtain. Knowledge/performance levels for tasks requiring expertise can be ranked using time-independent distribution functions. However, an individual through the intentional act of learning can over time move within a time-independent distribution. This movement is described by a time-dependent function called a "learning curve." Characteristic learning times can be derived from learning curves. However, the process of learning and its assessment are different for expertise than for the interpersonal. Traditionally, education focusses on teaching and assessing various areas of expertise. Learning along the interpersonal dimension is much more difficult to teach and assess for people and machines. However, as interpersonal tasks become of greater value economically, this dimension cannot be ignored.

Keywords Job tasks · Learning curves · Workplace automation · Ranking expertise · Characteristic learning times · Interpersonal job skills

3.1 The "Task Model" for Production

What exactly do people learn in order to be compensated in the labor market? The Bureau of Labor Statistics' *Occupational Outlook Handbook* that provided the data for the prior chapter on educational credentials and wages is organized by occupations—medical doctor, electrician, plumber, teacher, lawyer, scientist and so on. However, the day-to-day activities workers in these respective occupations perform have evolved in time, due in part to increased automation and computer usage. As a result, the daily work does not necessary reflect the educations and trainings the workers received. Skillsets individuals bring to the jobs they perform impact their wages, but because workers continue to learn, those skillsets also evolve in time.

© The Author(s), under exclusive license to Springer Nature Switzerland AG 2023   25
J. Ganem, *Understanding the Impact of Machine Learning on Labor and Education*,
SpringerBriefs in Philosophy, https://doi.org/10.1007/978-3-031-31004-1_3

Traditionally, jobs have been classified as white collar/blue collar or, in the labor market models referenced in the prior chapter, high-skilled/low-skilled. We think of white collar/high skill jobs as paying more than blue collar/low skill jobs. But many white-collar jobs do not require mastery of a particular specialized skill, while many blue-collar jobs do. The highly paid, white-collar construction project manager does not need the specialized skills that the lesser paid blue-collar construction workers— carpenters, masons, plumbers, electricians possess. It is also the case that many high-skilled jobs, such as artists pay little in comparison to many low-skilled jobs such as waiters.

To better understand the relationship between workers' educations, skillsets, and usage of computers, Autor et al. (2003) introduced a "task model" of production. In their model, the final output of a good/service is broken down into a sequence of discrete tasks performed by the workers. They sorted tasks into five categories— routine-manual, routine-cognitive, non-routine-manual, non-routine-cognitive, inter- personal. This last category refers to interactions with other people—communicating, managing, etc. that are inherently non-routine. The authors sought to understand how computerization altered the demand for various job skills. Their analysis of job task input over the years 1960–1998 found a measurable shift from routine tasks to non-routine and interpersonal tasks. They attributed this shift to computers either performing routine tasks or complementing worker performance of routine tasks. Their model explained much of the relative shift in demand to college-educated labor over this time period because college-educated workers have a comparative advantage in performing non-routine tasks.

Follow up analysis ten years later, used a job tasks model to show a predictive relationship between workers' skillsets and wages (Autor and Handel 2013). They modeled an occupation as an indivisible bundle of tasks to be performed simultane- ously by the worker. In contrast to an education, which is a static attribute possessed by a worker, tasks performed on the job "are an application of the worker's skill endowment to a given set of activities," and workers modify "these task inputs as job requirements change, which is an ongoing self-selection process consistent with comparative advantage." In other words, workers learn new job-related skills over time as the demands of the job change to stay relevant in the job market.

A later paper (Deming 2017) updated the original data analysis (Autor et al. 2003) to reflect the changing composition of job tasks over time. The analysis, now extended to the year 2012, showed a continuation of the trends found earlier—that is decreases in the performance of routine tasks and increases in tasks requiring "social skills" (Deming's term for "interpersonal" used in the earlier 2003 paper). He found a growing demand for social skills in the labor market because tasks that require workers to use these skills were continuing to increase at the expense of all other kinds of worker-performed tasks.

The reason Deming gives for the increased value on social skills is that teamwork enhances productivity through comparative advantage, as people with high social skills can seamlessly trade tasks with each other. Computers are poor at social skills and unable to enhance their productivity through interaction with others. He argues that workers with high social skills self-select into non-routine occupations, but he admits to not knowing where social skills come from and if they can be influenced by education or public policy.

3.2 Sorting Tasks into Categories

The task model is a compelling conceptualization of work and has shown to be a robust predictor of wages. However, the model raises two issues relevant to the expansion of machine learning into the workplace. (1) How are routine tasks differentiated from non-routine tasks? (2) Why are computers bad at interpersonal tasks?

In the creation of their task model, Autor et al. (2003) defined non-routine tasks in terms of Polanyi's description of "tacit knowledge," that is "tasks for which the rules are not sufficiently well understood to be specified in computer code and executed by machines." They quoted from Polanyi "We can know more than we can tell...." (Polanyi 1966, p. 4). "The skill of a driver cannot be replaced by a thorough schooling in the theory of the motorcar; the knowledge I have of my own body differs altogether from the knowledge of its physiology; and the rules of rhyming and prosody do not tell me what a poem told me, without any knowledge of its rules" (Polanyi 1966, p. 24).

Indeed, Autor et al. gave "navigating a car through city streets" as an example of a non-routine task by their definition. The reason is that that task, at the time of their writing, could not be accomplished using a programmable set of rules, even though the task is a "minor undertaking for most adults" (Autor et al. 2003). However, 15 years later, because of machine learning, computers can navigate a car through city streets, so in a sense the task could now be considered routine, even though the computer does not have an explicit step-by-step set of instructions for driving, which still makes it non-routine by their definition.

Machine learning has blurred the line between routine versus non-routine, but in the same sense that line is also blurred for humans. Referring to driving as a "minor undertaking" is another way of saying routine for most adults, but that is only because most adults *learn* to drive at some point in time. There is nothing routine about a person's first time driving a car. The same can be said for any activity that requires learning to perform—a task that is routine for the skilled practitioner is not routine for the novice.

The transformation from novice to expert provides a way to conceptualize learning, whether it is machine learning or human learning, *as a process unfolding in time in which rule-based knowledge becomes tacit knowledge*—that is non-routine tasks become routine tasks. In the labor market, the real value of expertise, be it in heart surgery or plumbing, is that tasks that are routine for the expert are so non-routine for non-experts, as to be impossible, which is why experts have comparative advantages.

But what is it about interpersonal tasks that always make them inherently non-routine? People skills such as listening, communicating, persuading, leading, and so on, often improve with age and experience, which means that these skills are learned. However, the learning of interpersonal skills does not transform interactions with people into routine tasks, which has been problematic for automating these tasks to an extent that allows machines to effectively perform them. Interpersonal and social skills are highly dependent on the situation and the people involved. For example, a particular individual can be a highly effective leader in one context and completely ineffective leader in a different context. In this sense, the situational dependence of interpersonal skills renders them different from expertise in plumbing, heart surgery, driving, and so on, which involve performing tasks independent from the social environment.

Therefore, this chapter will divide job tasks into two broad categories. (1) Those that use *expertise*—that is a standalone knowledgebase/skillset that can be independently assessed and ranked in relation to other individuals. (2) Those that are *interpersonal*—that is use a contextual social knowledgebase/skillset that is situationally dependent and cannot be independently accessed and ranked. Both knowledgebases/skillsets—expertise and interpersonal—are learned, which means that they are time-dependent personal attributes/traits/qualities with characteristic learning times (T_c). But conceptualizing and assessing the learning processes for the two categories is different.

3.3 Expertise

For the purposes of this chapter, an *expert* knowledgebase/skillset that an individual acquires through learning becomes an attribute, trait, or quality (Q) possessed by that individual. Learned Q's differ from other possible Q's in that they require time and agency to obtain. This chapter models learning times for tasks requiring expertise and formalizes the definitions of T_i and T_c introduced in Chap. 2.

3.3.1 Distribution Functions for Traits

Consider a large population of N individuals possessing an attribute, trait, or quality (Q). That attribute could be adult height, age, skill in playing chess, degree of fluency in Chinese and so on. Q varies between individuals and can be measured in terms of a numerical value x. For example, x could be adult height in inches, age in years, Elo rating in chess, score on a test of Chinese fluency, and so on. The values for x have a variance associated with them characterized by a parameter z. In other words, there is a distribution function, $f(x; z)$, such that:

$$n = \int_{x_1}^{x_2} f(x; z)dx \qquad (3.1)$$

where n is the number of individuals with a value of x between the values of x_1 and x_2 for the trait Q. The function $f(x; z)$ is normalized so that

$$\int_{-\infty}^{\infty} f(x; z)dx = N \qquad (3.2)$$

where N is the total number of individuals being assessed for the quality Q. It is also possible to compute an average x (labeled μ), associated with the quality Q.

$$\mu = \int_{-\infty}^{\infty} x[f(x; z)]dx \qquad (3.3)$$

For an individual's x_i, the total number of people n_i with an x less than x_i is

$$n_i = \int_{-\infty}^{x_i} f(x; z)dx \qquad (3.4)$$

Figure 3.1 shows an example f(x; z) and illustrates the general properties and parameters that a normalized distribution function for x must have.

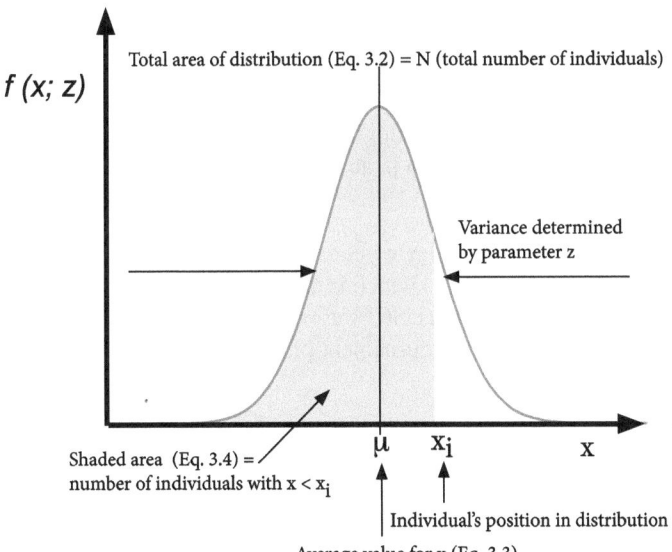

Fig. 3.1 Example of a distribution function $f(x; z)$ (Eq. 3.1) for an expertise Q. The value of x is a measure of Q

Therefore, the degree to which an individual possesses the quality Q relative to all others in the distribution can be expressed in terms of a percentile (P_i) given by:

$$P_i = 100\left(\frac{n_i}{N}\right) \tag{3.5}$$

3.3.2 Examples of Distributed Traits

Example 3.1 Let Q be adult female height in the United States. This is an example of an individual attribute Q, that cannot be learned but can be described with a distribution function. In this case, N = population of adult females in the United States, x = height and μ = average height. It is possible to model variability in height using a Gaussian or "normal" distribution function given by:

$$f(x; z) = \frac{N}{z\sqrt{2\pi}}e^{-\frac{(x-\mu)^2}{2z^2}} \tag{3.6}$$

where z is referred to as the "standard deviation" from the average height μ. Studies (Fryar et al. 2016) report that for adult females in the United States $\mu = 63.7$ inches and $z = 2.65$ inches. A woman with $x_i = 68$ inches tall—a typical height of a female fashion model—according to Eq. (3.5) with Eq. (3.6) for the distribution function, would be in the 95th percentile—meaning that she would be taller than 95% of the total population of adult females.

Note that for Q = adult height, if the population were expanded to include all adults, the distribution would not follow Eq. (3.6) exactly, but rather be a sum of two Eq. (3.6)'s (bi-model)—one with a μ and z for females and one with a different μ and z for males.

Example 3.2 Let Q be skill in playing chess measured in terms of x = player's Elo rating (Regan and Haworth 2011) which is an example of a learned attribute Q. The United States Chess Federation (USCF) uses a "logistic" distribution to model the variability in the Elo ratings of tournament players. A logistic distribution is given by:

$$f(x; z) = \frac{N}{2z}\text{sech}^2\left(\frac{x-\mu}{z}\right) \tag{3.7}$$

where N = the total number of rated players, μ = average Elo rating, z = variance in Elo ratings, and "sech" is the hyperbolic secant function. Ideally the distribution of ratings should be from 0 to 3000, with an average of 1500 and only 1% of players rated above 2200—the threshold for "master" status. This would require assigning players' ratings such that $\mu = 1500$ and $z = 320$ in the Eq. (3.7) formula.

In practice, the distribution of ratings among tournament players is much more complicated than Eq. (3.7). Even a bi-modal model—like the distribution of adult heights when men and women are included—does not work. The actual distribution of chess ratings is "multi-modal" because there are different populations of players—youth, adults, established players, new players—and each distinct population has different μ and z values for its distribution. This property of distributed traits (Q's)—that different populations have different averages and variances for a given Q—is common and means that in general, the function $f(x; z)$ in Eq. (3.1), can be complicated.

3.3.3 Conditions

For the purposes of the arguments in this chapter the following conditions are assumed when modeling the distribution of a particular trait Q:

(1) The exact mathematical form of $f(x; z)$ in Eq. (3.1) is not important—it can be symmetric like the distributions described by Eqs. (3.6 and 3.7), bi-model, multi-model, or skewed, and the variance z can be large or small.

(2) $f(x; z)$ must be a mathematical function that is "normalizable," meaning that the integral given by Eq. (3.2) must be finite, otherwise there would exist the absurdity of an infinite number of people in the population.

(3) $N \gg 1$, that is the total number of individuals possessing the trait is so large that the subtraction or addition of a single individual at any point along the x-axis does not result in any significant change to the distribution function $f(x; z)$. This assumption allows the use of continuous mathematics—x can be treated as a continuous variable—as opposed to discrete. This also allows for a single individual to move within the distribution without altering its overall functional form.

(4) The function $f(x; z)$ has no explicit time-dependence. Through birth and death, acquisition and attrition, individuals with the trait Q can enter or leave the distribution of Q at various points without changing the mathematical form of the distribution function or its average μ. It is easy to imagine a trait with a distribution function that evolves in time, but this is a separate problem that will not be considered in this chapter.

For example, if Q = SAT score for high school seniors, all four conditions above are met. (1) SAT scores are distributed nearly normally about an average $\mu = 1500$ so that Eq. (3.6) is a good model for $f(x; z)$ with z = 300. (2) The function described by Eq. (3.6) is normalizable. (3) There are millions of high school students that take the SAT, which means that the addition or subtraction of any one student with any given score has no effect on the overall distribution. (4) Even though each year a different population of students take the SAT, the test is intended to provide a standard such that the distribution of scores each year remains about the same—normally distributed about $\mu = 1500$ with z = 300.

Over long periods of time this final condition does break down for many traits. Average SAT scores have decreased in recent years. Over many decades the average height for adult females in the United States has increased. The distribution of other possible Q's, such as age also evolves slowly in time—life expectancy has slowly increased which has increased average age. But these are slow drifts and not relevant to the arguments in this chapter.

3.3.4   Time-Dependence of Q

While $f(x; z)$ does not change in time, an individual's x_i and its associated P_i (Eq. 3.5) for a given trait Q may or may not change over time. Because of Condition 3 above, x_i's can change in time without changing the overall distribution of the trait—given by $f(x; z)$. For example, a high school student could take the SAT, score 1200, take an SAT prep-course, and then re-take the SAT six months later and score 1500. Over the six months the student's x_i increased by 300, a change in percentile ranking from 16% to 50%, but the overall distribution (Eq. 3.6 with $z = 300$ and $\mu = 1500$) that included millions of students did not change.

In general, an individual's x_i can change in either direction or remain static. Its time derivative will be designated $g_i(t)$, which results in the differential equation:

$$\frac{dx_i}{dt} = g_i(t) \tag{3.8}$$

A change is possible for an x_i that specifies an individual's SAT score, but it is not possible for every Q an individual could possess. It is possible to sort Q's into three categories according to the allowed time-dependencies of their associated x_i's—(1) time-independent (2) unalterable time-dependent (3) variable time-dependent. For examples in each category consider:

Time-independent (Example—adult height)—an individual's adult height will not change significantly over a lifetime. This means that because of Condition 3, for a large population of N individuals, an individual's percentile ranking in the distribution function $f(x; z)$ for height does not change in time, that is $g_i(t) = 0$.

Unalterable time-dependence (Example—age)—an individual's age changes at a constant unalterable rate. For a large population, the entire distribution function is roughly fixed in time because of births and deaths, which satisfies Condition 3 above (we are ignoring the long-term drifts in life expectancies). However, an individual's position in the distribution function relative to all others does change at a constant rate because of the unavoidability of aging—that is the time derivative $g_i(t) =$ constant.

Variable time-dependence (Example chess playing skill)—an individual's skill at chess can change over time, which means that an individual chess player's position in the distribution function (x_i) and its associated P_i relative to the other chess players can change. In this case $g_i(t)$ can be a complicated function of time with negative values during some time periods (forgetting) and positive values during others (improving).

There are some learned Q's that are essentially time-independent. The special case of $g_i(t) = 0$ for a learned Q over long time periods corresponds to a knowledge-base/skillset that an individual neither forgets nor improves (e.g. riding a bicycle for most adults). It is also possible to have an unalterable time-dependent learned Q such that $g_i(t) =$ a positive constant. This corresponds to a knowledgebase/skillset under-going continuous improvement. There is no mathematical difference between this latter special case and an unalterable time-dependent Q such as age. The distinction between the two cases is that *a learned Q requires agency*, while a Q such as age does not because it cannot be learned.

Therefore, for a learned knowledgebase/skillset—e.g. chess—an individual's x_i and corresponding percentile ranking (P_i) *can* change in time. But "can" is a necessary qualification because unlike unlearned Q's with unalterable time-dependence—e.g. age—learned Q's are not compelled to change in time. In other words, an individual can choose to learn chess, or not to learn chess, or forget what was learned about chess in the past, but an individual cannot choose to avoid aging.

3.3.5 Time-Dependent Learning Curves

If x_i for a learned Q changes in time, the individual's percentile ranking (P_i) will change in time, but only within the bounds of 0 to 100 percent. That function $P_i(t)$ is referred to as the "learning curve." Figure 3.2 shows an example of a learning curve for an individual acquiring a particular quality Q, beginning as a complete novice $(P_i = 0$ at $t = 0)$ to complete mastery $(P_i = 100)$ as time progresses. The example curve has periods of increasing P_i (learning), periods of decreasing P_i (forgetting), and plateaus (no change in skill level).

Using a learning curve, a formal definition of T_i (individual learning time) will be made. On the learning curve locate the time T_- for the P_i given by:

$$P_i(T_-) = 100\left(\frac{1}{N}\right) \int_{-\infty}^{\mu-z} f(x; z)dx \tag{3.9}$$

That is the time at which the individual's $x_i = \mu - z$ (one degree of variance z less than average—see Fig. 3.2). Then locate the time T_+ corresponding to $x_i = \mu + z$ (one degree of variance z greater than average) given by:

$$P_i(T_+) = 100\left(\frac{1}{N}\right) \int_{-\infty}^{\mu+z} f(x; z)dx \tag{3.10}$$

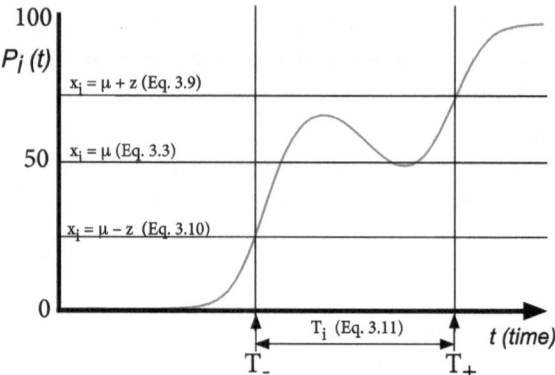

Fig. 3.2 Example of a learning curve for an individual. $P_i(t)$ is an individual's percentile ranking (between 0 and 100) within a distribution (Eq. 3.2) for an expertise Q as a function of time. An $x_i = \mu$ (average) equates to 50 percentile. The time interval between T_- (Eq. 3.9) and T_+ (Eq. 3.10) is defined as the individual learning time T_i (Eq. 3.11)

Then define:

$$T_i = T_+ - T_- \tag{3.11}$$

That is T_i is the time for the individual to improve in skill from one degree of variance below average to one degree of variance above average (see Fig. 3.2).[1]

For a large number ($N \gg 1$) of individuals learning a quality Q, each individual will have a unique learning curve $P_i(t)$ with an individual learning time (T_i). While learning curves apply to individuals, not populations, the T_i's will have a distribution function given by:

$$n = \int_{t_1}^{t_2} h(T_i; w)dt \tag{3.12}$$

where n is the number of individuals with an individual learning time T_i, between t_1 and t_2 and w is the variance for the T_i's. Then the characteristic learning time is defined as the average of T_i's, which is:

$$T_c = \int_0^{\infty} T_i[h(T_i; w)]dT_i \tag{3.13}$$

---

[1] Note that a "steep learning curve" would mean a short T_i—the individual learned quickly, which is the opposite of the common expression in English, which uses the phrase "steep learning curve" to mean difficult to learn.

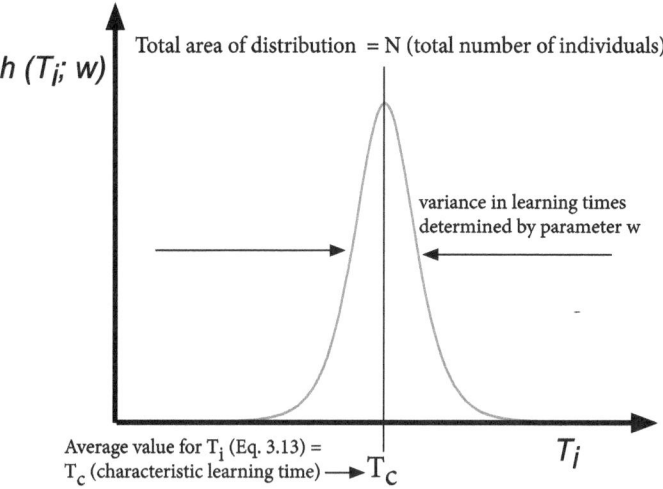

Fig. 3.3 Example of a distribution $h(T_i; w)$ for individual learning times T_i's for an expertise Q (Eq. 3.12). The characteristic learning time T_c is the average of the T_i's (Eq. 3.13)

In this expression the integral is over time, which means the limits must run from needing no time to learn ($T_i = 0$) to the indefinite future ($T_i = \infty$) because time can only go forward. Therefore, Eq. (3.13) defines the characteristic learning time (T_c) as an average of all the individual learning times (T_i's). Figure 3.3 shows an example distribution $h(T_i; w)$ for the T_i's.

Notice that the time scales in Figs. 3.2 and 3.3 could range by orders of magnitude—T_i could be measured in seconds or in decades. Unlike measures of expertise (P_i's), which are bounded between the 0 and 100 percentiles, learning times (T_i's) are not bounded. An individual could be a child chess prodigy and attain grandmaster status as a teenager, or study decades to become a grandmaster. Learning is not the same as knowing, which means that the learning curve $P_i(t)$ is not derivable from the distribution of expertise $f(x; z)$ *unless the function $g_i(t)$ is specified*. That is the time dependence of the learning process must be known. In addition, knowledge of an individual learning curve will not provide information on the distribution of individual learning times $h(T_i; w)$. Therefore, if $g_i(t)$ is specified it will apply only to an individual, which means that it will not be possible to derive the distribution $h(T_i; w)$.

However, there is a general relationship between the characteristic learning time scale (T_c) with its variance (w) and the variance in expertise (z). If the parameter z is large—meaning the distribution function $f(x; z)$ is wide—there is a large variance in expertise within the population of practitioners. In that case it is likely that the knowledgebase/skillset is difficult to learn and has a large T_c and w often measured in years (e.g. mastering chess). If the parameter z is small, that means that there is little variance in expertise within the population of practitioners. In that case it is likely that the knowledgebase/skillset is easy to learn and has a small T_c and w, which are often measured in hours (e.g. riding a bicycle). As argued in Chap. 2, expertise that can be learned in hours has much less economic value than expertise that requires years of learning.

3.3.6   Example Time-Dependent Learning Curves: Sigmoid Functions for Logistic Distributions

The time-dependence of an individual P_i (learning curve) for a quality Q, measured with a numerical value x_i that changes because of a learning process $g_i(t)$, can be illustrated with a specific example. Consider values for x distributed according to a logistic function (Eq. 3.7) for a large population of N individuals. Combining Eqs. 3.4, 3.5, and 3.7 results in:

$$P_i = \int_{-\infty}^{x_i} \frac{100}{2z} \operatorname{sech}^2\left(\frac{x - \mu}{z}\right) dx \tag{3.14}$$

If x_i has a time-dependence because of learning, then according to Eq. (3.14), P_i will be a function of time. Let us specify x_i in terms of the fractions (n) of variance (z) from the average (μ) and allow n to be a function of time. That is:

$$x_i = \mu + n(t)z \tag{3.15}$$

Substituting Eq. (3.15) for the upper limit x_i in Eq. (3.14) and integrating results in:

$$P_i(t) = 50(1 + \tanh[n(t)]) \tag{3.16}$$

where tanh is the hyperbolic tangent function. Now consider the function $g_i(t)$ (Eq. 3.8), which is obtained by differentiating Eq. (3.15).

$$\frac{dx_i}{dt} = g_i(t) = z\frac{dn(t)}{dt} \tag{3.17}$$

Solving for n(t) results in:

$$n(t) = \int_0^t \frac{1}{z} g_i(t') dt' \tag{3.18}$$

where t' is standard math notation for a "dummy variable" of integration. Therefore, by specifying a time-dependent learning process $g_i(t)$, we can integrate Eq. (3.18) and substitute the result into Eq. (3.16) to obtain the learning curve for an expertise Q distributed according to a logistic function.

For example, suppose the learner makes progress at a constant learning rate such that $g_i(t) = rz$, where r is a constant rate measured in fraction of variance (z) per time unit—which is the special case discussed in Sect. 3.3.4 in which $g_i(t) = $ constant. After substitution into Eq. (3.18) and integrating:

$$n(t) = rt + n_0 \qquad (3.19)$$

where n_o is the learner's starting point, measured in fractions of variance from the average. Therefore, for this constant rate learning process, in which the individual moves through a logistic distribution for expertise, the learning curve is:

$$P_i(t) = 50(1 + \text{Tanh}[(rt + n_0)]) \qquad (3.20)$$

This result (Eq. 3.20) is the well know "s-curve" or "sigmoid curve" that has been used for decades to model machine learning (Vapnik 2000). Figure 3.4 shows example sigmoid learning curves for three different individual learning rates $r = 0.03$, $r = 0.04$, and $r = 0.06$ all expressed in fraction of z per time unit. Using Eq. (3.11), these correspond to $T_i = 67$ time units, $T_i = 50$ time units, and $T_i = 33$ time units respectively. In Fig. 3.4 the learners all start with the same $n_0 = -3$ meaning 3z below the average μ, which corresponds to an initial $P_i = 0.25\%$ (almost no prior knowledgebase/skillset). The three curves show comparative progress for the three learners over 200 units of time. Note that a time unit could be hours, days, weeks, months, years. These learning curves would scale according to whatever unit of time is used. If T_i's were determined from individual learning curves for a large population of learners, the distribution of T_i's would result in $h(T_i; w)$. From that distribution, Eq. (3.13) could be used to calculate T_c.

3.4 Interpersonal

Expertise, however sophisticated and advanced, will always be subject to automation, because to be an "expert" means that difficult tasks are now routine—that is automatic for the expert. If these tasks are automatic for an expert, they can in principle be automated by a machine. Indeed, computers have upended the economics of many industries—law, accounting, journalism, publishing, architecture for examples—because inexpensive software packages perform tasks—drafting standard legal documents, preparing tax returns, writing sports summaries, designing book layouts, drawing floor plans—that in the past required a highly trained expert.

As articulated at the beginning of this chapter, becoming an expert is a learning process in which explicit rule-based knowledge is over time transformed into tacit knowledge. Experts can be ranked on a distribution curve relative to one another because the end goals are stable even if the knowledgebases/skillsets have changed

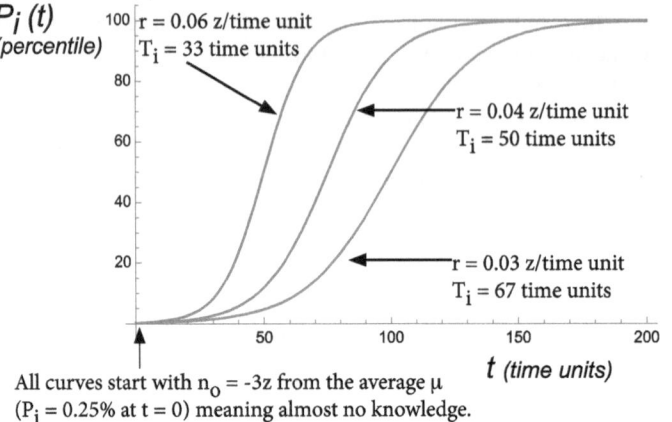

All curves start with $n_0 = -3z$ from the average μ
($P_i = 0.25\%$ at $t = 0$) meaning almost no knowledge.

Fig. 3.4 Example of sigmoid learning curves Eq. (3.20) for an expertise Q with a logistic distribution (Eq. 3.7) for three different learning rates r (Eq. 3.19). The curves are for $r = 0.03$, $r = 0.04$, and $r = 0.06$ all expressed in fraction of z per time unit. Using Eq. (3.14), these correspond to $T_i = 67$ time units, $T_i = 50$ time units, and $T_i = 33$ time units respectively. The learners all start with the same initial knowledgebase/skillset ($n_0 = -3$)

in time. The techniques and tools for a heart bypass operation have changed in time but the underlying pathology being treated is still the same. Plumbers use different tools and materials, but a clogged pipe is fundamentally the same problem that it has always been. Chess theory has evolved substantially over time, but the object of the game—checkmate—has not changed.

However, interpersonal interactions have no stable end points. Collegial, political, personal, and family relationships begin with chance encounters, evolve over time and that evolution is driven by circumstances and intentions. In some cases, participants in a relationship intentionally severe it, but more common is for the participants to "move on" to some other job, project, location, stage-in-life, and the relationship is renegotiated or becomes dormant. A relationship is successful when there is reciprocity that serves mutual needs. A relationship ends or lapses into dormancy when circumstances change such that reciprocity can no longer be maintained.

In James P. Carse's book, *Finite and Infinite Games* (Carse 1986), he divides activities in life—"games" in his terminology—into two kinds: "finite games" played for the purpose of "winning" and "infinite games" played for the purpose of continuing play. Finite games by must have a specified beginning, explicit rules, and end when all players agree on the "winner." While there can be only one winner of a finite game, the contestants can be ranked at the conclusion of play. In contrast, an "infinite game" has no beginning or end. The entire purpose of an infinite game is to prevent it from ending, to keep everyone in play.

To use Carse's paradigm in the labor market, a finite game would be competing for and obtaining a specific job with a specific job title. An infinite game would be sustaining a career. A finite game would be performing a heart bypass operation on a single patient. An infinite game would be improving the practice of cardiology. A finite game would be unclogging the kitchen sink on a particular day at a specific address. An infinite game would be establishing and growing a plumbing services company.

Finite games involve expertise that can be ranked relative to others, which means that the analysis in the prior chapter using distribution functions and percentiles is applicable. This is possible because finite games have stable endpoints and can be repeated. Through study and repetition finite players learn and improve their rankings. However, infinite players cannot be ranked because without a stable endpoint there is no repetition. Work requiring "non-routine" interpersonal job tasks epitomizes play in an infinite game. Yet, infinite games in the labor market almost always require a learned—that is time-dependent—interpersonal knowledgebase/skillset. This presents challenging problems when educating humans and machines. What are the desired learning outcomes for mastering an infinite game when there is a single criterion for success in such an endeavour—the continuation of play? How can "success" even be defined without an endpoint?

Traditionally, education has ignored these questions by focusing on expertise—teaching the knowledgebases/skillsets that allow for individuals to be measured and ranked using distribution functions for various areas of expertise. But labor market analysis using job tasks models shows that expertise is becoming less important in determining wages. The book *The Robot Factory: Pseudoscience in Education and Its Threat to American Democracy* (Ganem 2018), notes a paradox in twenty-first century education practices and outcomes. "Educated people are becoming increasingly better off, while at the same time, much of the valuable work traditionally performed by educated people is being offloaded to computers." This implies that there are additional education outcomes beyond acquiring expert knowledgebases/skillsets that can be evaluated and ranked. The job task data documenting the increase in non-routine tasks requiring interpersonal/social skills as the performance of routine tasks decline would explain this observation, given that educated workers tend to have comparative advantages for non-routine tasks.

It is becoming clear that in the modern workplace, knowing/performing is not the entire source of value. Managers, for example, do not need the skillsets of the people they manage but they do need to judge performance levels and know the levels (x_i's) necessary to achieve the outcomes. They need to communicate and coordinate individual job tasks to achieve goals that no one person can accomplish alone. Most importantly, the long-term strategy needs to be set. There needs to be a motivation for the work and reasons for sustaining an "infinite game."

At this point in time, machines are finite game players. Even machines with sophisticated artificial intelligence and machine learning algorithms exist to perform specific defined tasks. It is true that machines have been created to simulate human relationships (Reis et al. 2020). But a machine created to provide human companionship, or even intimacy (Devlin 2018), is still playing a finite game in that it will be judged on its ability to relate to a human and discarded if it cannot meet the human needs and expectations. A machine cannot be a player in an infinite game because it cannot choose to play. As Carse notes (Carse 1986), players in an infinite game must *choose* to do so.

But as artificial intelligence and machine learning evolves, machines will take on interpersonal tasks in the future. How do we educate and assess machines on interpersonal knowledge and skills without the ability to rank them, given that we don't know how to educate and rank humans' interpersonal abilities? This question must be faced because in the future machine intelligence, in addition to performing tasks requiring expertise, will also take on interpersonal tasks, or at a minimum complement human interpersonal work.

Machines might also develop intelligences beyond the categories of expertise and interpersonal discussed in this chapter. To date, work in artificial intelligence has focused on replicating human intelligence. But this is an anthropocentric view of intelligence. Why not, for example, replicate dog intelligence? Dogs after all possess intelligence in regard to navigating the world by scent that humans lack, but humans find valuable and therefore employ dogs to perform. It is also possible that in the future machines will develop superintelligences—that is intelligences that humans cannot replicate and might not even understand. Managing and interacting with such an alien artificial intelligence would require a machine with "interpersonal" skills to mediate between human intelligence and various superintelligences. After all, how would a human judge the effectiveness and performance of a superintelligence that humans cannot understand?

As new tasks are created, this distinguishability problem—judging the usefulness and effectiveness of an alien artificial intelligence—becomes an interpersonal skill itself. A managerial intelligence might not need a superintelligence in one specialized area, but it will need enough specialized knowledge to judge the validity of inputs from competing sources. In fact, the judgment problem might be of greater importance than the knowing/performing problem. To understand this difference, Turing's "Imitation Game" will be revisited in Chap. 4.

3.5 Summary

A job requires the performance of a bundled sequence of discrete *learned* tasks. The required tasks can be sorted into those needing *expertise* and those needing *interpersonal* skills. The bundled tasks for a given job are a mix of expertise and interpersonal. However, the mix is not necessarily equal. Some jobs might have mostly interpersonal tasks and some jobs mostly tasks requiring expertise. Learned knowledgebases/skillsets along both dimensions—expertise and interpersonal—for a particular job require agency and time to obtain. However, when labor is bought and sold in the marketplace, the participants value knowing/performing.

An *expert* knowledgebase/skillset can be considered an individual attribute (Q) and can be independently assessed and ranked relative to other individuals using a distribution function $f(x; z)$. Q's that individuals learn have individual learning times (T_i's) associated with them. But the distribution function $h(T_i; w)$ for the T_i's is different than, and independent from, the distribution function $f(x; z)$ for the attribute itself Q. The distribution of x_i's for Q provides no information on the

(T_i's) that the various individuals needed to acquire their x_i's and percentile rankings (P_i's). Learning is fundamentally different than knowing/performing because learning has a time-dependence while knowing/performing does not. The act of knowing/performing takes place at a point in time, while learning unfolds in time. Stating that an individual performs a skill at a specific level—say 75th percentile relative to the entire population—provides no information as to how long it took that individual to achieve that performance level, or the present time rate of change of that performance level. Was that 75th percentile performance achieved by an individual learning a new skill very rapidly, or an individual with a declining skill level who performed in the 90th percentile in the past? The performance level itself does not provide that temporal information.

However, the x_i's are what determine the value of the labor. It is the skill of the heart surgeon or plumber relative to others with a comparable skillset that matters. How long it took a practitioner to acquire his or her skillset is not relevant to the immediate transaction. But, as Chap. 2 established, the wage premiums that some occupations have over others depends on a characteristic learning time T_c [Eq. (3.13)] and choice of career for an individual depends on his or her T_i [Eq. (3.11)] and its relationship to T_c. Therefore, that temporal information is of vital economic importance. In addition, because expertise by its nature—defined tasks with stable endpoints—can in principle be automated, the economic impact of machine learning on the market for expertise will also depend on how fast machines can learn—not just what they are capable of learning.

An individual's interpersonal knowledgebase/skillset is learned, but it is difficult to *independently* rank and assess because of its situational dependence. Job tasks requiring interpersonal skills are inherently non-routine and stable endpoints do not exist. But interpersonal tasks are an increasing component of most jobs, while expert tasks are declining. Effective performance of interpersonal tasks results in increased wages, and a comparative advantage for non-routine tasks. Employers can and do judge and rank interpersonal skillsets/knowledgebases. Managers, for example, routinely consider a worker's interpersonal skillset, in addition to the worker's expertise when assigning job tasks and evaluating overall job performance. However, the teaching, assessing, and ranking of interpersonal skillsets has been more of a byproduct, rather than an explicit intent, of an education system focused on teaching expertise. But as the market value of interpersonal tasks increases and machine learning reaches a level that allows computers to complement humans, or even substitute for them, in the performance of interpersonal tasks, this judgment problem becomes of great importance.

References

Autor DH, Levy F, Murnane RJ (2003) The skill content of recent technological change: an empirical exploration. Q J Econ 118(4):1279–1333. https://EconPapers.repec.org/RePEc:oup:qjecon:v: 118:y:2003:i:4:p:1279-1333

Autor DH, Handel MJ (2013) Putting tasks to the test: human capital, job tasks, and wages. J Law Econ 31(S1):S59–S96. https://doi.org/10.1086/669332

Carse JP (1986) Finite and infinite games: a vision of life as play and possibility. The Free Press

Deming DJ (2017) The growing importance of social skills in the labor market. Quart J Econ 132(4):1593–1640. https://doi.org/10.1093/qje/qjx022

Devlin K (2018) Turned on: science, sex and robots. Bloomsbury Sigman

Fryar CD, Gu Q, Ogden CL, Flegal KM (2016) Anthropometric reference data for children and adults: United States, 2011–2014. National Center for Health Statistics. Vital Health Stat 3(39). https://www.cdc.gov/nchs/data/series/sr_03/sr03_039.pdf

Ganem J (2018) The robot factory: pseudoscience in education and its threat to American democracy. Springer International.https://doi.org/10.1007/978-3-319-77860-0

Polanyi M (1966) The tacit dimension. Doubleday Press

Regan KW, Haworth GM (2011) Intrinsic chess ratings. In: Proceedings of the twenty-fifth AAAI conference on artificial intelligence, vol 25, pp 834–839. https://doi.org/10.1609/aaai.v25i1. 7951

Reis J, Melao N, Salvadorinho J, Soares B, Rosete A (2020) Service robots in the hospitality industry: the case of Henn-na hotel, Japan. Technol Soc 6:101423. https://doi.org/10.1016/j.tec hsoc.2020.101423

Vapnik VN (2000) The nature of statistical learning theory, 2nd edn. Springer

Chapter 4
The Judgment Game: The Turing Test as a General Research Framework

Abstract Turing's three-participant "Imitation Game," is revisited and the probabilistic and temporal nature of the game is formalized. It is argued that Turing-like games can be used as general tests of distinguishability between levels of knowledge/performance in learned subject areas along the expertise and interpersonal dimensions. A modification of the Imitation Game called the "Judgment Game," is introduced, which also has three participants—an interrogator (Judge), who must distinguish between two players A & B. The players, despite inherent differences, attempt to be indistinguishable to the Judge. However, in the Judgment Game, the focus is on the success of the Judge, which can be scored and compared to other judges. Because the Judge's "job" is mostly interpersonal, such a game provides a method to assess knowledgebases/skillsets, along the interpersonal dimension, which has been problematic in the education of humans and machines. A machine in the role of the Judge in a Judgement Game, might be a higher standard for machine cognition than a machine in the role of Player A, as in Turing's original Imitation Game. However, these tests provide a restrictive definition of "thinking" because thought processes involved with learning must take place before these games can be played.

Keywords Turing test · Imitation game · Judgment game · Turing test as a research framework · Limitations of the Turing test · Machine cognition standards

For humans, performing—that is the use of an existing knowledge base/skillset, be it expertise or interpersonal knowledge, and learning—that is the acquisition of new knowledge—are both activities that require thinking. Therefore, a test such as Turing's "Imitation Game" that defines machine thinking, does not necessarily distinguish between these two different thought processes—performing versus learning. The preceding Chap. 3 argued that an inherent difference between performing versus learning is the element of time. A performance takes place at a particular point in time, while learning takes place over time. With this distinction in mind, consider the elements of Turing's game and prior analyses of it.

The "imitation game" is "played with three people, a man (A), a woman (B), and an interrogator (C) who may be of either sex." The interrogator stays in a room apart

J. Ganem, *Understanding the Impact of Machine Learning on Labor and Education*, SpringerBriefs in Philosophy, https://doi.org/10.1007/978-3-031-31004-1_4

from the other two and asks A and B questions, in writing, so that there is exists no sensory contact—visual or voice communication—between them. The object of the game for the interrogator is to determine which is the man and which is the woman. A and B are obligated to answer the questions but have different goals. For A, the object of the game is to cause the interrogator to make the wrong identification. For B, the object is to help the interrogator make the correct identification. At the end of the game, the interrogator must state which person is the man, who is trying to deceive, and which person is the woman, who is trying to help. Turing then imagines replacing person A, the man who is trying to deceive, with a machine. Will the interrogator make the wrong identification with a machine in the role of A, as often as with a man in the role of A? Turing proposes that this question replace the question: "Can machines think?" (Turing 1950) This thought experiment has come to be known as the "Turing test."

4.1   Observations on the "Imitation Game"

1. *The Imitation Game is a performance test because it takes place at a specific point in time.* As such, it defines "thinking" as the use of existing knowledge-bases/skillsets. Thought processes that result in learning (or forgetting), thus modifying the existing knowledgebases/skillsets, are not demonstrated/tested in the Imitation Game.
2. *There are two interpretations of the test—species versus gender.* There is a dispute in the literature as to whether Turing intended that the interrogator be misled into believing that A is another human being or be misled into believing that A is a woman as opposed to a man. Sterrett (2000) argues that a literal reading of Turing's paper shows that he describes two tests for machine intelligence, and that while he regarded them as equivalent, she contends that the two tests are not. She designates the tests as "The Original Imitation Game Test," and "Standard Turing Test." The Original Test is the one Turing presents first, described above, in which both man and machine must impersonate a woman. Later in Chap. 5 of the paper Turing writes: "Let us fix our attention on one particular digital computer C. Is it true that by modifying this computer to have an adequate storage, suitably increasing its speed of action, and providing it with an appropriate program, C can be made to play satisfactorily the part of A in the Imitation Game, the part of B being taken by a man?" Sterrett argues that this is a different test because now only the computer is attempting to impersonate, that is make itself indistinguishable from a man.

   This second test is the usual understanding of the Turing test—that the computer make itself indistinguishable from a human being. But as Genova (1994) has pointed out this is a "species test" not the "gender test" that Turing first describes. Sterret concurs with Genova and goes on to argue that The Original Imitation Game Test is more meaningful because it is less dependent on the skill of the interrogator whereas passing The Standard Turing test is highly dependent on the skill of the interrogator.

In response, Moore (2001) argues against Sterret's gender interpretation of the test. He observes that throughout the paper, Turing focuses on tasks that require human intellectual ability and he contends that Turing's use of the word "man" is in a generic sense. Moore writes: "Turing focuses upon humans as a group and seeks to compare differences between humans and machines, not women and machines or women and men." He also disagrees with Sterret's contention that a gender test is a better test of machine intelligence because he argues that the "gender test" can be embedded into the "standard test," in addition to other types of skills such as creating poetry or designing a house. However, Moore does acknowledge that the skills, knowledge, and even biases of the interrogators are an important part of the imitation game. He discusses interrogation strategies at length and the qualities of a "good Turing question."

3. *Most AI researchers do not take the Turing test seriously.* Since 1991, an annual competition for the Loebner Prize (Wikipedia 2021a) has pitted computer programs against each other in attempts to pass the Standard Turing Test. Human judges score each program's responses to decide which are most closely human. While no program has ever passed the Standard Turing Test, the prize is awarded to the program with the highest score. However, many AI researchers dismiss this competition as a publicity stunt because it focusses on creating programs that mimic conversation so that an interrogator will have difficulty discerning if a conversation is with a person or a machine.

In contrast, most AI research is focused on solving economically important problems—not trying to fool another human into thinking that a machine is human. AI researchers want machines with superhuman abilities for narrow and specific tasks—e.g. reading X-rays, driving cars. No one has interest in machines that mask their strengths at calculation, as Turing proposed in his paper, as a way of mimicking the slowness of humans in arithmetic in order to trick a human. Some AI researchers have argued that the goal of passing the Turing Test is even harmful to the field of AI because it is based on an unnecessarily restrictive view of intelligence. Hayes and Ford (1995) write that: "All versions of the Turing Test are based on a massively anthropocentric view of the nature of intelligence." They write:

> From a practical perspective, why would anyone want to build machines that could pass the Turing Test? As many have observed, there is no shortage of humans, and we already have well-proven ways of making more of them. Human cognition, even high-quality human cognition, is not in short supply. What extra functionality would such a machine provide, even if we could build it? (Hayes and Ford 1995)

In addition, like Moor (2001) and Sterret (2000), Hayes and Ford (1995) acknowledge that the "… the success of the game depends crucially on how clever, knowledgeable, and insightful the judge is. As a result: "The imitation game will not have a stable endpoint." In other words, because it is possible for the judge to improve at the game, the standard that the game sets for "thinking" could increase over time."

4. *The Turing test can be regarded as a general research procedure.* Moore points out that "Turing himself treats the imitation game both as a general research technique modifiable and applicable to various problems and as the now famous test of human impersonation given by the standard interpretation." Because Turing described more than one version of the imitation game, Moore argues that for "Turing the imitation game is a format for judges to impartially compare and evaluate outputs from different systems while ignoring the source of the outputs." Indeed, other researchers have used "Turing-like" tests for such purposes. For example, Colby (1981) used Turing-like indistinguishability tests to evaluate the ability of clinical judges to distinguish model paranoid patients from actual paranoid patients.

5. *The outcome of this game is binary*—that is the interrogator must state either A or B. Statements such as "I am 90% confident in A" or "I am 60% confident in A" are not part of the test. Nor is the element of time mentioned. However, Turing did think of this game as probabilistic and temporal. He provided a probabilistic and temporal definition of success—if the machine fooled an "average" interrogator 70% of the time after *5 minutes* of questioning it would be considered as thinking. Implicit in this statistical definition of success is that the interrogator possesses some level of skill that can be ranked in relation to an "average" and that there must be repeated trials of the game until a statistically significant success rate is established. It should be noted, however, that this version of the game—repeated trials with binary outcomes—is different than asking the interrogator to make a choice and then state a confidence level.

For example, suppose the interrogator made no attempt to distinguish the man from the machine and simply flipped a coin. That would result in a guaranteed 50% success rate. However, if the interrogator made a good faith effort to judge but only achieved a 50% success rate, that result would be indistinguishable from coin-flipping. If the interrogator was easily duped—say 20% success rate—a large improvement in the interrogator's success rate could be achieved by resorting to coin-flipping. This allows for the possibility of learning on the part of the interrogator. Turing's qualifier of "average" implies that the role of interrogator requires some level of skill. But, as already mentioned in Observation 3, repeated trials of a 5-minute question and answer might not converge to a stable average success rate if the interrogator is learning while doing and improving as time progresses.

In addition, the outcome of the game can vary depending on whether the interrogator *knows* that a substitution has been made. If the interrogator is told that one participant is a machine and the other is a person, coin-flipping guarantees a minimum 50% success rate. But if random substitutions are made—two people become a machine and a person, or two machines, without the interrogator being informed—the interrogator must make a different kind of judgment and the 50%

baseline rate for success is no longer assured. For example, can players in an online game detect when a machine has been suddenly substituted for a human player? Without that information being provided, they might not notice or even consider the possibility that they are now playing against a machine instead of a human (Pfau et al. 2020).

4.2 The Judgment Game: A Modified Imitation Game to Rate the Skill of the Interrogator

This book uses the "Imitation Game" format as a general research framework, which means that Turing's intentions are not relevant. It will conceptualize the Imitation Game as a generalized test for indistinguishability, in which Players A & B are considered inputs and the interrogator's ruling is a binary true/false output signal. If the average output is not significantly different statistically than random coin flipping, then Players A & B are indistinguishable *based on the criteria used by the interrogator* because the interrogator must use criteria for making a judgment. If not, the output cannot be anything other than coin flipping—it defaults to randomness.

It will also default to randomness if Players A & B are, in fact, indistinguishable and no *valid* criteria can be constructed. Suppose one of the players claims to be better at roulette than the other player. In a game of pure chance, no *valid* criteria can be constructed that will distinguish Player A from Player B, so the claim is not testable. No such thing as "expertise" exists in a game of pure chance like roulette. This means that the Players A & B must, in principle, be distinguishable from one another for such a Turing-like test to be meaningful, but they will not be distinguished if the interrogator's criteria are insufficient or faulty.

Therefore, consider a modified Imitation Game that will be termed "The Judgement Game." The question this game poses: Does the interrogator have the ability (criteria plus skills) to differentiate two players (A & B) with different percentile rankings in a specified area of expertise, or would the two players be indistinguishable from the point of view of the interrogator? Again, the expertise must be real and have sufficient depth that a meaningful distribution function for the expertise can be constructed. It cannot be imagined—like asserting that skill exists in a game of chance—or trivial—like the game of tic-tac-toe.

The Judgment Game is played with three human participants designated as the "Imposter" (Player A) the "Expert" (Player B) and the "Judge" (interrogator). The goal of the game is for the Judge to identify the "Imposter." In the setup of the game, a specialized area of knowledge accessible through learning is announced, and a person who possesses a measured level of learned knowledge/skillset in that area— given by a percentile ranking (P_i)—is designated the "Expert." The "Imposter" is a person claiming to have that same level of knowledge/skillset, but in fact has less. The Judge must interrogate both persons and determine which one is the actual expert. The Judge is informed at the beginning that only one person is the Expert.

For examples, the Expert could be a master-level chess player ($P_{i = Expert} = 99\%$) and the Imposter a novice ($P_{i = Imposter} = 10\%$) or the Expert a native, but not highly educated, Chinese speaker $P_{i = Expert} = 75\%$ and the Imposter a person who took four years of Chinese-language classes in college $P_{i = Imposter} = 20\%$. Both chess and Chinese are learned with levels of expertise that can be ranked with P_i's from distribution functions. To use the terminology of Chap. 3, the knowledgebase/skillset is a specific Q (a learned attribute = chess, Chinese .... etc.). The game is set up so that the percentile ranking on the distribution function for the Expert is always greater than that for the Imposter—that is $P_{i = Expert} > P_{i = Imposter}$. The Imposter's goal is to become indistinguishable from the Expert in the assessment of the Judge.

The setup of the Judgment Game is somewhat like the American TV Show: *To Tell the Truth*, in which a panel of judges must distinguish an "expert" in a specified occupation from two imposters (Wikipedia 2021b). In, *To Tell the Truth*, as in Turing's original Imitation Game, the imposters are permitted to engage in intentional deception—such as lying to the judge(s). However, the Judgment Game, like the Imitation Game and unlike *To Tell the Truth*, has just one Imposter, and one Judge.

However, in the Judgment Game it is not the success of the imposter that is of interest, as it is on the TV show and in the Imitation Game, but rather the success of the Judge. The Judge's performance can be scored by assigning a point value (s_n) of +1-point for a correct identification and −1-point for a wrong identification for the nth trial. The Judge's total score (s_{Judge}) after $N = \Sigma$ n trials, will be defined as:

$$s_{Judge} = 10 \cdot \sum s_n/N \qquad (4.1)$$

A savvy Judge, who can always correctly identify the imposter, has an $s_{Judge} = 10$ over many trials. It is possible that an easily duped Judge can have an $s_{Judge} < 0$, although an s_{Judge} of at least 0 with this evaluation system can always be obtained by coin-flipping. If $s_{Judge} = 0$ the Imposter has succeeded in becoming indistinguishable from the Expert in the assessment of the Judge. However, a Judge that succeeded 70% of the time (Turing's definition of success) would have $s_{Judge} = 4$, which will be designated as the "Turing threshold."

Figure 4.1 provides a schematic description of the Judgment Game. A constraint in the Judgment Game is that no learning takes place. The level of expertise for the Imposter and the Expert is fixed and the Judge uses static criteria in making a binary ruling. The Judge either correctly identifies the imposter $s_n = +1$ or does not $s_n = -1$. If the Judge's performance (s_{Judge}) converges to a value less than 4, then the Imposter and the Expert are indistinguishable according to the Judge's criteria, even though they are in principle distinguishable. Therefore, the effectiveness of the Judge's criteria and skill can be ranked along the scale shown in Fig. 4.1 from −10 to +10, where −10 is assigned to a Judge who is always wrong, 0 is indistinguishable from random guessing and +10 is assigned to a Judge who is always correct.

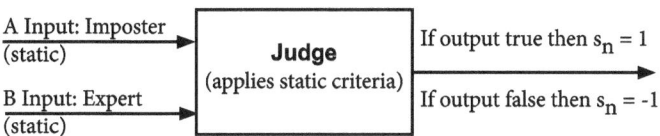

(a) Judgment Game

A: Imposter ≠ B: Expert
(distinguishable in principle)

A Input: Imposter
(static)

B Input: Expert
(static)

Judge
(applies static criteria)

If output true then $s_n = 1$

If output false then $s_n = -1$

Repeat until to total number of trials $N = \Sigma n$ is statistically signficant. Then $s_{Judge} = 10 \cdot \Sigma s_n / N$.

(b) Ranking a Judge

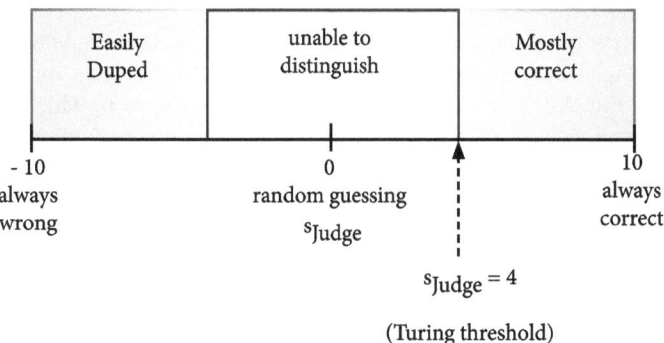

| Easily Duped | unable to distinguish | Mostly correct |

- 10
always
wrong

0
random guessing
s_{Judge}

10
always
correct

$s_{Judge} = 4$

(Turing threshold)

Fig. 4.1 Schematics of **a** Judgment Game, **b** scoring system for judges (Eq. 4.1)

4.3   The Job of the Judge

If we think of the participants in the Judgment Game has having jobs, it is useful to consider the bundle of tasks each is assigned. Each participant's job requires the performance of interpersonal and expert tasks. All participants must use interpersonal skills to communicate with each other and all must have expertise if relevant questions are to be formulated, answered, and understood. The area of expertise could be very general—the expertise needed for two human beings to converse with each other in a way that makes mutual sense (the Standard Turing Test). Or the area of expertise could be very narrow—playing high-level chess. But the conversation must have a topic requiring some level of expertise in some area, or else it would be gibberish for all involved.

However, the Judge's job, and the bundle of tasks it involves, is different than the jobs and the bundle of tasks required for the two players A & B. The players need only to communicate their level of expertise in response to relevant questions. Their job tasks are weighted heavily to those requiring expertise because their goal is to demonstrate expertise. The Judge's job is weighted heavily towards tasks requiring interpersonal skills because the Judge is acting much like a manager. To formulate meaningful questions, the Judge must have some level of expertise but not necessarily at the level of players A & B. Instead, the Judge must rely on interpersonal skills to communicate with A & B, ask good questions, and discern the differences in A & B's levels of expertise based on the answers they provide. That is $P_{i\,=\,Judge} > 0$ or the Judge would have no basis for formulating relevant questions, let alone making judgements. But the Judge must also possess the interpersonal skillset necessary to ask good questions and evaluate the answers.

Therefore, the score for the Judge will depend on the levels of expertise for all three of the participants in addition to the Judge's interpersonal ranking—designated IP. That is the Judge's score is:

$$s_{Judge} = S\left[P_{i=Expert}, P_{i=Imposter}, P_{i=Judge}, IP_{Judge}\right] \tag{4.2}$$

where the function S is a complicated function of the four inputs given.

Consider Fig. 4.2 that shows possible distributions of scores s_{Judge} for a population of judges. In Fig. 4.2a, the distribution is centered at $s_{Judge} = 0$, which means that for the average judge, A & B are indistinguishable, and there are almost no judges above the $s_{Judge} = 4$ Turing threshold. For the distribution in Fig. 4.2b, the average judge is above the Turing threshold and only a small minority of the judges fail in distinguishing A from B. Therefore, it is distributions like the one shown in Fig. 4.2b that can be used to rank the effectiveness of judges.

Of course, the difficulty of the Judge's task will depend on the difference in the level of expertise between the Expert and the Imposter. For example, distinguishing a novice chess player $P_i = 10\%$ from the average tournament player $P_i = 50\%$ is not that difficult a task and would require a modest level of chess expertise on the part of the Judge. However, it would be very hard to distinguish a chess master $P_i = 99\%$ from a chess grandmaster $P_i = 99.99\%$ and would require that the Judge have a great deal of chess expertise. Similar issues would occur with languages—easy to distinguish a first-year student of Chinese from a native speaker, but difficult to rank two native Chinese speakers on their respective knowledge of the language.

However, such a game could provide a way of ranking interpersonal skills, which so far has confounded the education system. Because $P_{i=Judge}$ can be measured beforehand, such a game could be used to score a Judge's interpersonal skillset (IP) when the P_i's for all three players are held fixed. For example, suppose it is found that some, but not all judges, with a $P_{i\,=\,Judge} = 5\%$ for chess can reach the Turing threshold of $s_{Judge} = 4$ given the task of distinguishing a master-level chess player $P_{i\,=\,Expert} = 99\%$ from a novice $P_{i\,=\,Imposter} = 10\%$ resulting in a distribution for s_{Judge} scores like Fig. 4.2b. This implies that the score for each judge depends on his or her interpersonal skills (IP) and the distribution of s_{Judge} scores could serve as a proxy for assigning IP rankings to judges using percentiles.

For such an assessment to work, a $P_{i\,=\,Judge}$ would need to be found that results in a distribution of s_{Judge} scores. If $P_{i\,=\,Judge} = 0$, then $s_{Judge} = 0$ for all judges because judges with no knowledge would have no basis for making assessments. But if $P_{i\,=\,Judge} > 90\%$ then $s_{Judge} = 10$ for all judges because any judge with that level of

(a) A & B are indistinguishable - same as random guessing

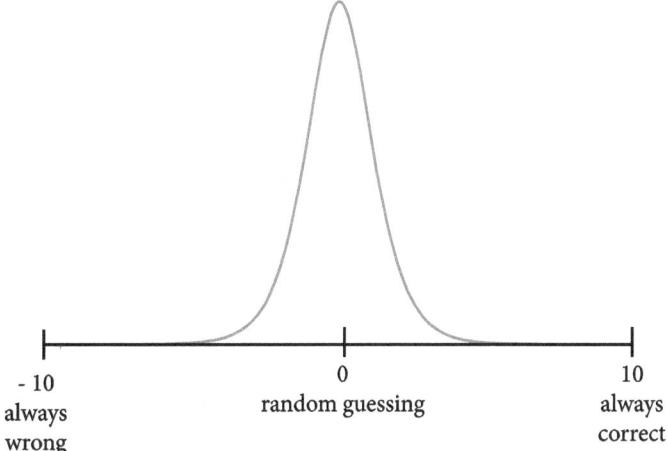

(b) A & B are distinguishable - above Turing threshold

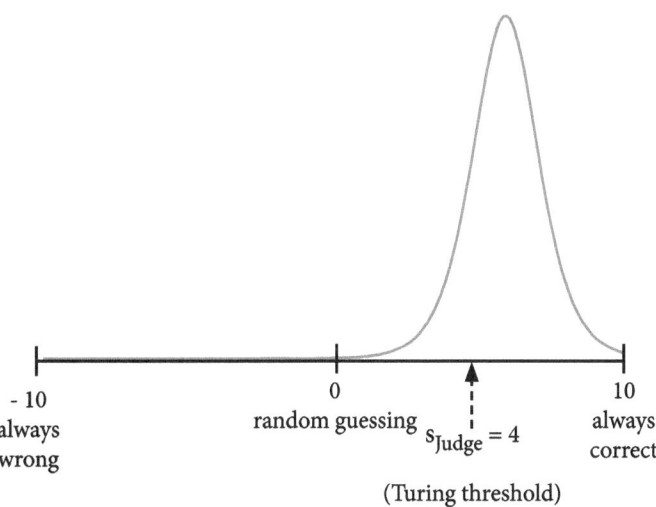

Fig. 4.2 Example distributions of s_{judge} scores (Eq. 4.1) for a large population of judges playing a Judgment Game. **a** An average judge cannot distinguish between players A & B (random guessing), **b** an average judge can distinguish between players A & B more than 70% of the time (above the Turing threshold)

expertise would always be able to distinguish a novice chess player from a master regardless of his or her IP. It follows that there must be some value for $P_{i = Judge}$, such that scores are distributed according to the interpersonal skill levels of the judges in asking questions and interpreting answers.

In fact, a game like this could be used to determine the minimum level of expertise needed by an "average" Judge to achieve the Turing threshold on assessments. Of course, that minimum level of expertise depends on the relative rankings of the "Expert" and "Imposter." As just stated, it takes a great deal of chess expertise to distinguish a chess master from a chess grandmaster, but not much expertise to distinguish a grandmaster from a novice.

We could also recast the "Standard Turing Test" and the "Original Imitation Game" using this Judgment Game format. When this is done, the entire game becomes a test of interpersonal skillsets because no "expertise" as we have defined it is involved. It is not possible to assess and rank "human" or "gender" using a distribution function.

For example, consider a Judgment Game set up like the original Imitation Game (without the computer substitution for the man). The players are: Expert = woman (Player B), Imposter = man (Player A), Judge = human of either gender (interrogator). The goal is to determine the s_{Judge} for the Judge in differentiating the man impersonating the woman from the woman. From a job task perspective, the skillset the Judge must use is entirely interpersonal, because gender is not an attribute that can be assessed and ranked using a distribution function in the ways that chess-expertise or Chinese-fluency can be ranked. Gender identification does appear to exist along a spectrum, with some people identifying as strongly male, some as strongly female, and some as non-binary. But that distribution of gender identification is not an "expertise" that can be ranked because there is no *inherent* value added or subtracted by a person's gender identification.[1] Being in the 95th percentile in the distribution for chess ratings denotes expertise that has economic value; being in the 95th percentile for "gender" is a meaningless statement. However, some judges will be better at the task of identifying imposters than others and the distribution of s_{Judge} scores for a population of judges could serve as proxy for measuring IP.

Similarly, imagine a Judgment Game set up like a Standard Turing Test: Expert = human (Player B), Imposter = machine (Player A), Judge = human (interrogator). To date, no machine has passed the Standard Turing test, meaning that $s_{Judge} = 10$ for all reasonably competent judges. However, if machines do approach the "Turing Threshold" where $s_{Judge} = 4$ for reasonably competent judges, it will not be a sudden change, but rather a gradual shift in the s_{Judge} distribution over time as machines become better at impersonating humans. In addition, the s_{Judge} scores will have a distribution during the transition to $s_{Judge} = 0$ (complete indistinguishability) because some humans will be better than other humans, to varying degrees, at making the judgment while this transition is underway.

As pointed out, passing the Standard Turing test isn't all that interesting a problem from an economic point of view. But what might be an economically interesting problem is a Judgment Game set up as follows: Expert = human with expertise in a specified area, Imposter = human novice in the same specified area, Judge = machine. The question becomes: How does the s_{Judge} score for a machine compare to the s_{Judge} score for an average human judge? For narrow areas of expertise, machines

---

[1] On average, men earn significantly more money than women, but this is a cultural artifact that has nothing to do with the argument being made here.

in many cases are already as good as humans. A machine can assess human chess-expertise, Chinese-fluency, and so on with a reasonable degree of accuracy. In some areas, machines might be even better than humans—financial advising, radiology.

But what if the job tasks involved are interpersonal? For example, imagine a machine playing the part of the interrogator in the Original Imitation Game or the Standard Turing Test. Would a machine have the interpersonal skills needed to distinguish a woman from a man impersonating a woman, or a human from a machine impersonating a human? What would the machine's s_{Judge} score be? The Judgment Game, with a machine in the role as Judge, could provide a conceptual framework for assessing a machine's IP for job tasks that are primarily interpersonal.

4.4 Summary

Turing-like tests, such as the "Imitation Game," the "Standard Turing Test," and the "Judgment Game," can be used as general tests of distinguishability between levels of expertise in learned subject areas. However, these kinds of tests evaluate performance; they do not evaluate learning. Therefore, these tests provide a restrictive definition of "thinking" because only thought processes that involve possession of a prior, existing knowledgebase/skillset are demonstrated. The thought processes involved with adaptive intelligence—learning—must have already taken place before these games can even be played. In addition, it is possible for learning and forgetting to take place in the future, which means that these tests do not provide a stable standard to define "thinking."

In all these games, the knowledgebase/skillset of the interrogator is a critical component in the outcome. The success of Player A in becoming indistinguishable from Player B depends on the difference in skill levels between Player A and Player B *and* the interrogator's skills in asking questions and interpreting answers. Because the results of these tests are always dependent on the skill of the interrogator, it is possible to hold the inputs fixed (Players A & B knowledgebases/skillsets) and use the results from the game to measure the skill level involved in the role of interrogator.

The actions and goals of all the players can be framed as "jobs" as defined by economic models for the labor market. If the players' roles in these games are considered jobs, the job tasks for the players include those that require knowledge of the subject matter (expertise) and the ability to communicate effectively (interpersonal). However, the interrogator's tasks are heavily weighted towards those requiring an interpersonal knowledgebase/skillset. The interrogator does not necessarily need a high level of expertise to distinguish Player A from Player B. The interrogator needs just enough expertise to formulate good questions and make sound judgments. Therefore, the Judgment Game could in principle be used to rate the effectiveness of the Judge's interpersonal skillset in a specific context—an assessment problem that has confounded educators who have traditionally focused on teaching and assessing expertise. However, because labor market trends show that job tasks requiring expertise are being displaced by interpersonal tasks, educators need to think better about how to teach and assess interpersonal skills. In addition, AI researchers who want to build a machine with an interpersonal skillset could use a Judgment Game to assess its effectiveness with the machine in the role of Judge.

References

Colby KM (1981) Modeling a paranoid mind. Behav Brain Sci 4(4):515–534. https://doi.org/10.1017/S0140525X00000030

Genova J (1994) Turing's sexual guessing game. Soc Epistemol 8(4):313–326. https://doi.org/10.1080/02691729408578758

Hayes P, Ford K (1995) Turing test considered harmful. In: Proceedings of the fourteenth international joint conference on artificial intelligence, pp 972–997. https://www.ijcai.org/Proceedings/95-1/Papers/125.pdf

Moor JH (2001) The status and future of the Turing test. Mind Mach 11:77–93. https://doi.org/10.1023/A:1011218925467

Pfau J, Smeddinck JD, Bikas I, Malaka R (2020) Bot or not? User perceptions of player substitution with deep player behavior models. In: Proceedings of the 2020 CHI conference on human factors in computing systems, pp 1–10. https://doi.org/10.1145/3313831.3376223

Sterrett SG (2000) Turing's two tests for intelligence. Mind Mach 10:541–559. https://doi.org/10.1023/A:1011242120015

Turing AM (1950) Computer machinery and intelligence. Mind LIX:433–460. https://doi.org/10.1093/mind/LIX.236.433

Wikipedia (2021a) Loebner Prize. Retrieved 20 Nov 2021, from https://en.wikipedia.org/wiki/Loebner_Prize

Wikipedia (2021b) To tell the truth. Retrieved 20 Nov 2021, from https://en.wikipedia.org/wiki/To_Tell_the_Truth

Chapter 5
The Learning Game: A Time-Dependent Turing Test

Abstract A time-dependent version of the Judgment Game called the "Learning Game" is presented. It permits two of the three participants to learn. The Judge is permitted to learn along the expertise and interpersonal dimensions, and Player A is permitted to learn along the expertise dimension, while the knowledgebases/skillsets of Player B are kept fixed to establish a comparison standard. In this game, the Judge's ability to distinguish between Players A & B becomes a complicated function of time, which means that different Learning Games can be constructed to simulate different learning environments. Three special cases are considered: (1) An environment much like a typical education setting, in which the "Judge" is a teacher whose job is teach Player A (student) to become indistinguishable from Player B (graduate). (2) An environment in which the "Judge" is a manager whose job is to direct employees with much greater expertise. (3) An environment in which the "Judge" is a manager whose job is to direct employees with much less expertise. It is possible to substitute a machine for either Player A or the Judge and use a Learning Game to train the machine and assess its performance.

Keywords Time-dependent Turing test · Learning Game · Training machines · Assessing machine performance · Assessing education outcomes · Management training

In the previous chapter on the Imitation Game, and variations of it such as the Judgment Game, it was noted that (1) these games demonstrate thought processes for performing, not learning; (2) the interrogator must use criteria in making a judgment, or else Players A & B are indistinguishable by default; (3) the interpersonal skillset of the interrogator plays a crucial role in the games' outcomes; (4) it is possible to set up these games with a machine in the role of interrogator; (5) these games do not have stable end points if the level of expertise and/or the interpersonal skills possessed by the interrogator change in time.

It is this last point—the time dependence—that is of interest in this chapter. For a Judgment Game, the outcome—s_{Judge} for the Judge—depends on the levels of expertise for the three players plus the interpersonal skillset (IP) of the Judge—Eq. (4.2). If these inputs are held constant, then s_{Judge} for the Judge must be constant. Now

© The Author(s), under exclusive license to Springer Nature Switzerland AG 2023   55
J. Ganem, *Understanding the Impact of Machine Learning on Labor and Education*,
SpringerBriefs in Philosophy, https://doi.org/10.1007/978-3-031-31004-1_5

consider a variation of the Judgment Game that will be termed the "Learning Game," in which the inputs have time dependencies, meaning that learning (or forgetting) takes place. In the arguments that follow, the $P_{i\,=\,Expert}$ will be held fixed in time so that the Expert functions as a "standard" while the P_i for the Imposter and the P_i and IP for the Judge are allowed to vary in time. As a result, the s_{Judge} score will vary in time, that is the function S defined by Eq. (4.2) now has time dependence according to:

$$S_{Judge}(t) = S\left[P_{i=\,Expert}, P_{i=\,Imposter}(t), P_{i=\,Judge}(t), IP_{Judge}(t)\right] \quad (5.1)$$

As a result, the s_{Judge} score for the Judge has a time-dependence given by Eq. (5.1). In general, if two of the three participants are allowed to learn, and the Judge is allowed to learn along two dimensions—expertise and interpersonal—then $s_{Judge}(t)$ becomes a complicated function of time. Figure 5.1 depicts the Learning Game schematically and an example $s_{Judge}(t)$ function. However, it is useful to consider three special limiting cases for Eq. (5.1) and their implications.

5.1 Limiting Cases of the Learning Game

5.1.1 Limiting Case 1—Education

Consider a Learning Game set up with the following conditions:

$P_{i\,=\,Judge} > P_{i\,=\,Expert}$ = time-independent learning goal(s);
$IP_{Judge} = 50$ percentile (constant) as determined by a Judgment Game (Sect. 4.2);
$P_{i\,=\,Imposter}(t)$, with $P_{i\,=\,Imposter}(t = 0) = 0$ as an initial condition.

Under these conditions, the only time-dependence for Eq. (5.1) arises from the function $P_{i\,=\,Imposter}(t)$. This limiting case is equivalent to a typical education setting in which the teacher is a Judge, the Expert is a graduate (person that has achieved the established learning goal(s) in the area of knowledge being tested), and the Imposter is a student. The Judge (teacher) has average and constant interpersonal skills (IP's). For example, the Expert could be a prior student who passed the final exam in a course that the Imposter (new student) is just beginning.

Under these conditions $S_{Judge} = 10$ initially because it is obvious to the teacher that the student is a beginner. However, at the end of the course, if the student is indistinguishable from any other student that has met the same learning goal(s), then $s_{Judge} = 0$ because the teacher (Judge) can no longer differentiate between the student (Imposter) and a graduate (Expert). In this limiting case the point of the game is to determine the function $s_{Judge}(t)$ that results from the learning process. Figure 5.2 shows a schematic of Limiting Case 1 and an example $s(t)$ that could result. Because only the student is allowed to learn, the time-dependence in $s_{Judge}(t)$ arises only from the student's progress. That is

Learning Game
General Case
A ≠ B
(distinguishable in principle)

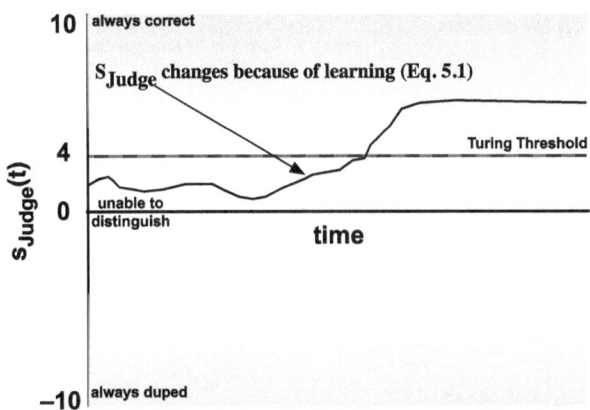

Fig. 5.1 Schematic of a Learning Game general case. S_{Judge} is a function of time (Eq. 5.1) because of learning by the Imposter along the expertise dimension, and the Judge along the expertise and interpersonal dimensions

$$S_{Judge}(t) = S\left[P_{i=Expert}, P_{i=Imposter}(t), P_{i=Judge}, IP_{Judge}\right] \qquad (5.2)$$

If we define a function:

$$L(t) = 100 - 10\, S_{Judge}(t) \qquad (5.3)$$

That function will be a learning curve function like the example in Fig. 3.2 in Chap. 3—beginning at 0 percentile (no knowledge) and transitioning to a plateau at 100 percentile (mastery of the learning goal). Notice that although it is the score of the Judge (teacher) that is changing, it is because the Imposter (student) is learning—not the Judge. Such a Learning Game would provide a T_i for the student in that subject area (Eq. 3.11). If played by many students, the average of the T_i's would provide T_c (Eq. 3.13). Note giving the final every day to measure progress in time toward meeting the learning goals is different than administering a final once at the end of a defined time-period (semester), which is how most education programs

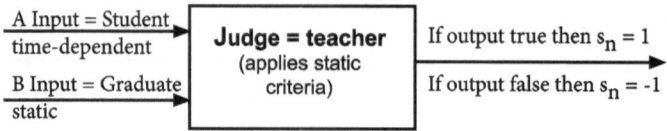

Learning Game
Case 1: Education
$$A \neq B$$
(distinguishable in principle)

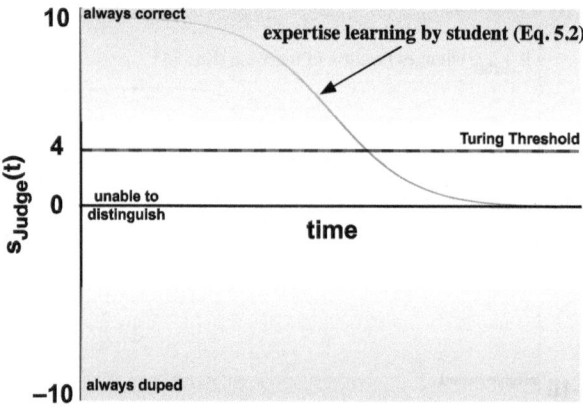

Fig. 5.2 Schematic of a Learning Game depicting Limiting Case 1—Education. S_{Judge} is a function of time (Eq. 5.2) because of learning by the Imposter

are structured. Learning curves cannot be determined from the final grade data in a traditional education program. This is because traditional education prepares students for a labor market in which wages are based on performing (not learning), so only performing is assessed. For example, an A on a transcript for a course that required Alice to study ten hours per week to achieve is indistinguishable from the A on Bob's transcript for the same course, even if he spent 20 hours per week studying to achieve the same learning goals. The different T_i's for Alice and Bob are not reported because in traditional education settings it is performance that counts, not learning time.

Now imagine substituting machines with learning algorithms for the Imposter (student). The question: How does T_c for the machines compare to T_c for humans? In the terminology of Chap. 2: Does the machine have a comparative learning advantage?

There are three cases to consider:

1. $T_{c\,=\,\text{machines}} \ll T_{c\,=\,\text{humans}}$ For this case, machines have a significant comparative learning advantage and if the task if economically important, there is no point to a human learning it. Consider, for examples, the games of chess and Go. Machines can now *learn* to master these games in a small fraction of the time it would take a human to become a master. Chess and Go are played for entertainment, so this fact about machines is of no economic importance. However, if these games and other similar games did have economic value, learning to master them would certainly be offloaded to machines. Likewise, humans should not spend time mastering any job tasks of economic value that a machine can master with a learning-algorithm in a fraction of the time it would take a human.

2. $T_{c\,=\,\text{machines}} \sim T_{c\,=\,\text{humans}}$ For this case, humans and machines can learn the job tasks in roughly equal spans of time. Neither has a comparative learning advantage over the other *at a given point in time*. However, time scales (T_c's) for human learning are biologically limited and cannot be changed significantly, while machine learning, in principle, has no inherent limitation on T_c. That means, going forward, machines can only reduce their T_c's, while humans cannot. Therefore, and this point is of central consideration in job retraining, humans should not spend time learning a new skill if a machine can learn to master the same skill in a roughly equal amount of time. If this is the case, the relevant time (T_r) on the job market for the newly learned skill will be too short for the human to recover opportunity costs, let alone profit from the retraining.

3. $T_{c\,=\,\text{machines}} \gg T_{c\,=\,\text{humans}}$ For this case humans have a comparative learning advantage that might persist for a significant time into the future. An example would be language translation. Machine learning algorithms, such as *Google Translate*, provide a clunky, superficial rendering, of text in one language into another language. The algorithm has difficulty reproducing much of the nuance, subtlety, and context found in natural language that a human translator, with an understanding of languages, culture, and social context would be able to capture. Google translate might help high school students cheat on assignments in their foreign language classes, but it won't be filling in at the United Nations for human translators any time soon. Expertise like language translation, that humans can still learn faster and better than machines, are ones that will be valuable in the labor market because of this comparative learning advantage.

5.1.2 Limiting Case 2—Management Training

Consider a Learning Game set up with the following conditions:

$P_{i\,=\,\text{Imposter}} < P_{i\,=\,\text{Expert}} = $ time-independent standard;
$IP_{\text{Judge}} = 50$ percentile (constant) as determined by a Judgment Game (Sect. 4.2);
$P_{i\,=\,\text{Judge}}(t)$, with $P_{i\,=\,\text{Judge}}(t = 0) = 0$ as an initial condition.

Under these conditions, the only time-dependence for Eq. (5.1) arises from the function $P_{i=Judge}(t)$. That is

$$S_{Judge}(t) = S[P_{i=Expert}, P_{i=Imposter}, P_{i=Judge}(t), IP_{Judge}] \tag{5.4}$$

The Judge's goal is to learn enough to distinguish the Expert from the Imposter. Figure 5.3 is a schematic of Limiting Case 2 with an example $s_{Judge}(t)$ function in which a Judge transitions from being unable to distinguish standard from substandard work at the beginning to always being able to distinguish at the end. For this case the change in s_{Judge} over time is the result of learning along the expertise dimension by the Judge. Note, the Judge does not need to achieve the expert's level of skill/knowledge, in contrast to the Judge (teacher) in Limiting Case 1 who needs to know more than the Expert (graduating student). In fact, in the end the Judge might still know less than the Imposter but have enough knowledge to distinguish standard from substandard work.

Learning Game
Case 2: Management Training
A ≠ B
(distinguishable in principle)

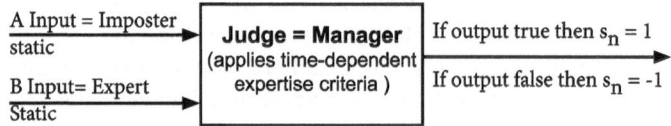

Repeat at equal time intervals until $s_{Judge}(t) = 10 \cdot \Sigma\, s_n / N$ approaches 10 (non-expert Manager can always differentiate between and Expert and Imposter).

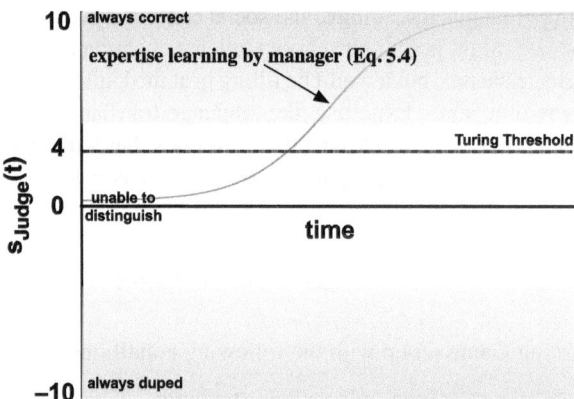

Fig. 5.3 Schematic of a Learning Game depicting Limiting Case 2—Management Training. S_{Judge} is a function of time (Eq. 5.4) because of learning by the Judge along the expertise dimension

This limiting case is typical of the management problem, in which the Judge is the manager who must direct workers who often have far more expertise in producing the saleable goods/services. As a result, while management is essential it is also paradoxical. Experts are too narrowed and focused to see the entirety of a company's vision and goals. But managers striving to execute a larger plan are stymied if they have no expertise. This tension provides endless comedic material for the comic strip *Dilbert*, in which the "pointy-haired boss" is clueless regarding the work his engineers actually do. The fact that this comic strip resonates so well is evidence that this management problem is universal.

A Learning Game played for the purpose of training the Judge could provide a framework for determining the level of expertise needed to manage and its T_c. Like the Judgment Game, the higher the standard and the smaller the difference between standard set by the Expert and the substandard performance of the Imposter, the more difficult it is to be the Judge—more expertise required and a longer T_c.

Could a machine be the Judge in Limiting Case 2? This is a relevant question for imaging a future in which machines manage. However, Limiting Case 2 is set up to assume an average interpersonal skillset (IP)—a level machines have yet to achieve—and focusses on learning expertise—something that machines in many areas of knowledge already have.

5.1.3 Limiting Case 3—Interpersonal Training

Consider a Learning Game set up with the following conditions:

$P_{i\,=\,Imposter} < P_{i\,=\,Judge} = P_{i\,=\,Expert} =$ time-independent standard;
$IP_{Judge}(t)$, with $IP_{Judge}(t = 0) = 0$ (no interpersonal skills) as an initial condition.

Under these conditions, the only time-dependence for Eq. (5.1) arises from the function $IP_{Judge}(t)$. That is

$$S_{Judge}(t) = S\big[P_{i=Expert}, P_{i=Imposter}, P_{i=Judge}, IP_{Judge}(t)\big] \qquad (5.5)$$

The Judge's goal is to acquire the interpersonal skills needed to distinguish the Expert from the Imposter. In this case, the Judge has the same knowledge as the Expert but initially cannot identify the Imposter because of a lack of interpersonal skills. Figure 5.4 is a schematic of Limiting Case 3 with an example $s_{Judge}(t)$ function in which a Judge transitions from being unable to distinguish the Expert from the Imposter to always being able to distinguish. The example $s_{Judge}(t)$ function is like the one in Fig. 5.3, but it arises for a different reason. For Limiting Case 3, the change in s_{Judge} over time is the result of learning along the *interpersonal* dimension by the Judge (not the expertise dimension as in Limiting Case 2).

For example, imagine a Standard Turing Test as performed in the Loebner Competition referenced in Chap. 4 (Wikipedia 2021). Both the Judge and the Expert are human, and the Imposter is a machine. The Judge and Expert are equal in terms of

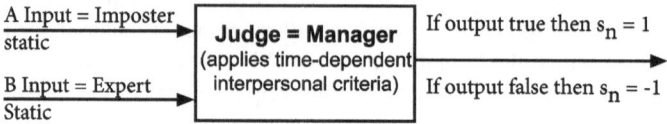

Learning Game
Case 3: Interpersonal Training
A ≠ B
(distinguishable in principle)

A Input = Imposter
static

Judge = Manager
(applies time-dependent
interpersonal criteria)

If output true then $s_n = 1$

B Input = Expert
Static

If output false then $s_n = -1$

Repeat at equal time intervals until $s_{judge}(t) = 10 \cdot \Sigma\, s_n /$
N approaches 10 (expert Manager can always differentiate
between and Expert and Imposter).

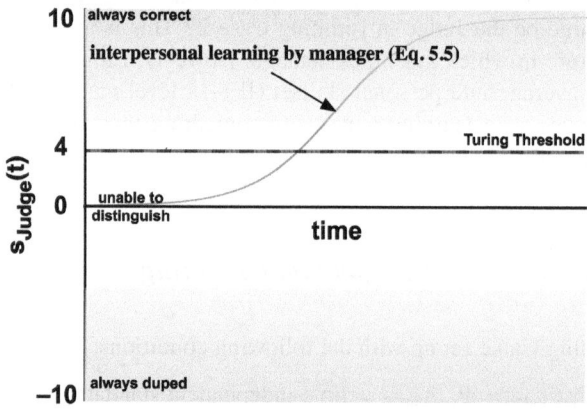

Fig. 5.4 Schematic of a Learning Game depicting Limiting Case 3—Interpersonal Training. S$_{Judge}$ is a function of time (Eq. 5.5) because of learning by the Judge along the interpersonal dimension

human cognition, but the Judge does not have the interpersonal skillset, matched to this situation, to ask good "Turing questions" and is fooled by the machine. Has the machine passed the Turing test? In the prior chapter on the Judgment Game, it was argued that if this experiment were performed with many different judges, and some succeeded in identifying the machine while others failed, the distribution that results could be used as a proxy for ranking IP's and could give meaning to Turing's use of the phrase "average interrogator."

But, instead of a Judgment Game with many judges, imagine a single Judge playing a Learning Game. If it were possible for a Judge to *learn* over time to identify the computer by developing an interpersonal skillset matched to this situation, then machine "thinking" is still identifiably different than human cognition. If it is not possible, then the Turing test is passed in such a way that a stable standard for "thinking" has been achieved. That is the quality of the "Turing questions" are no longer a factor in the standard.

Learning Games with the starting conditions of Limiting Case 3 could be played to develop an interpersonal skillset needed for judgment in any area of knowledge. For example, a complement to the "Management Problem" in Limiting Case 2, is the "Expert Problem," that occurs when a person, with a high level of expertise but limited interpersonal skills, is called on to manage. A Learning Game could be constructed to develop the interpersonal skillset needed to manage. Given that the education system teaches expertise, but job tasks are increasingly interpersonal, integrating Learning Games into education might help cross this divide between education goals and workplace demands. Chapter 3 noted that studies showing the increasing importance of interpersonal skills in the workplace have explicitly admitted a lack of knowledge on how interpersonal skills are learned and if the education system can better teach these skills (Deming 2017). Learning Games setup with the starting parameters of Limiting Case 3 could be a method for teaching and assessing interpersonal skill sets.

A Limiting Case 3 setup for a Learning Game could also be played with a machine in the role of Judge. If machines become managers in the future, they will need to learn an interpersonal skillset to be effective. Limiting Case 3 could become a testing framework for assessing and improving the interpersonal skills of machines. In fact, an original Imitation Game with a machine in the role of the interrogator attempting to distinguish a woman from a man pretending to be a woman might be an even higher standard for machine cognition than the original game in which the interrogator is always human.

5.2 Summary

Turing's Imitation Game has had the flaw that it is not a stable standard. However, that instability can be exploited as a test for learning. The Imitation Game can be used as a general research framework for testing the ability of a "Judge" to distinguish between two inputs. That ability, which depends on the skill and knowledge of the Judge, becomes a "performance" when the skillset and knowledgebase of all the participants are fixed. This time-independent use of an Imitation Game to assess performance is called the "Judgement Game" in Chap. 4. However, when the participants in the game are allowed to learn (or forget), that is their skillsets and knowledgebases change in time, the outcome of the game also changes in time. This time-dependent version of the Judgement Game is called the "Learning Game."

Chapter 3 established that learning takes place along two dimensions—expertise and interpersonal. AI research and human education programs have generally focused on expertise, which involves tasks with stable endpoints. The stability of the final goal means that tasks can be repeated, and individual performances ranked using a distribution function, which makes assessment straightforward. Interpersonal knowledge is usually just assumed to be present and is often treated as a constant. Desired education outcomes, for example, are almost always specified in terms of ranked expertise—learning goals for math, reading, science, etc.—that are usually defined with a distribution function without any reference to the interpersonal skills of the students, the teachers, or the administrators.

But the Learning Game can be used to assess both dimensions of learning because the roles (jobs) the players are assigned require different skillsets/knowledgebases. There can be no game unless all the participants have some minimal level of inter-personal and expert knowledgebases/skillsets. However, the Judge's role is primarily interpersonal while the Expert/Imposter roles rely primarily on expertise. This means that output of the Learning Game is in general a complicated function of time as knowledgebases/skillsets for all three participants change in time along both dimen-sions—interpersonal and expertise. It is also possible to imagine a machine substi-tuting for a human in anyone of the three roles in the Learning Game to assess if a machine would be more or less effective than a human for learning a specific job.

The complexity of the Learning Game output is reduced when only one of the participants is allowed to learn along just one of the knowledgebase/skillset dimensions. Three cases are of special interest:

1. *The Imposter is allowed to learn along the expertise dimension while the Expert and Judge's knowledgebases/skillsets along both dimensions are held fixed* (Fig. 5.2). This approximates a typical education setting in which Judge = teacher, Imposter = student, and Expert = graduate. In this limiting case the time-dependent Learning Game output can be converted to learning curves (see Chap. 3) and used to determine T_c for the specific area of expertise taught. By substituting a machine for the imposter, it is possible to compare the T_c for the machine with the T_c for the human to determine which one has a comparative learning advantage (see Chap. 2).

2. *The Judge is allowed to learn along the expertise dimension while the Imposter and Expert's knowledgebases/skillsets along both dimensions are held fixed* (Fig. 5.3). This approximates a typical management setting in which Judge = manager, Imposter = substandard worker, and Expert = standard worker. In this version of the game, the Judge lacks expertise but has the interpersonal skillset needed to manage. The goal is for the Judge to acquire enough expertise to evaluate the abilities and effectiveness of the workers.

3. *The Judge is allowed to learn along the interpersonal dimension while the Imposter's and Expert's knowledgebases/skillsets along both dimensions are held fixed* (Fig. 5.4). This approximates the "expert problem" in which Judge = manager, Imposter = substandard worker, and Expert = standard worker. In this version of the game, the Judge is an expert that lacks the interpersonal skillset to manage. The goal in this game is to train the Judge on interpersonal skills. Such a setup could also be used to train and assess the effectiveness of a machine in a management role.

These are three particularly relevant examples of many possible Learning Games that could be constructed for a given set of initial conditions and learning goals along the two dimensions—expertise and interpersonal. Other Learning Games could have additional uses in training and assessment of humans and machines for various job tasks. These could be especially useful for learning and assessing tasks along the interpersonal dimension, which has proved problematic in assessing humans in traditional education settings, and especially difficult in assessing the performance of machines that must engage in interpersonal interactions with humans.

References

Deming DJ (2017) The growing importance of social skills in the labor market. Q J Econ 132(4):1593–1640. https://doi.org/10.1093/qje/qjx022
Wikipedia (2021) Loebner Prize. Retrieved 20 Nov 2021, from https://en.wikipedia.org/wiki/Loebner_Prize

Chapter 6
Implications: Recommendations for Future Education and Labor Policies

Abstract A job in which there is a lack of time-dependence for the Judge's score can be modeled as Judgement Game, while the existence of a time-dependence for the Judge's score means that the job must be modeled as a Learning Game. Humans have a comparative learning advantage in Learning Games, while learning-enabled machines have a comparative learning advantage in Judgment Games. Humans should focus on jobs that can be modeled as Learning Games. Current education practices need revision because they focus on teaching knowledgebases/skillsets used in jobs that resemble Judgment Games. This contributes to the current mismatch in the labor market—in which many people are looking for work while, at the same time, many jobs are going unfilled. As a remedy, counterintuitive recommendations for changes in labor and education policies are provided. These include a renewed focus on liberal arts education rather than the current trend that emphasizes technical training, and that employers should screen job candidates for learning ability rather than specific skillsets. In addition, if AI is to work *with* humans, it must have knowledgebases/skillsets along the expertise *and* interpersonal dimensions. The framework presented provides a practical means for separating the interpersonal dimension from the expertise dimension.

Keywords Advantages of a liberal arts education · Future job training programs · Labor market changes resulting from artificial intelligence · Moravec's Paradox · Computer chess · Computer poker

> … it has become clear that it is comparatively easy to make computers exhibit adult-level performance in solving problems on intelligence tests or playing checkers, and difficult or impossible to give them the skills of a one-year-old when it comes to perception and mobility.
>
> — Hans Moravec (1988)

This quote is an articulation of what has become known as "Moravec's Paradox." It is the observation by early AI researchers that the knowledge-based recall and reasoning skills, that humans spend years in schools working to master, requires relatively little in terms of computational resources. However, the sensory knowledge/skills—psychomotor abilities, spatial perception, and emotional awareness—that humans take for granted and devote relatively little time to deliberate formal learning, require enormous computational resources. As linguist and cognitive scientist Steven Pinker writes in his 1994 book *The Language Instinct*:

> The main lesson of thirty-five years of AI research is that the hard problems are easy and the easy problems are hard. The mental abilities of a four-year-old that we take for granted – recognizing a face, lifting a pencil, walking across a room, answering a question – in fact solve some of the hardest engineering problems ever conceived... As the new generation of intelligent devices appears, it will be the stock analysts and petrochemical engineers and parole board members who are in danger of being replaced by machines. The gardeners, receptionists, and cooks are secure in their jobs for decades to come. (Pinker 1994)

To use the language of this book—machines have a comparative learning advantage for knowledge-based recall and reasoning, while humans have a comparative learning advantage for sensory and psychomotor skills. However, the fact that knowledge-based recall and reasoning is difficult for humans has, in the past, made these kinds of performance abilities economically valuable—a straightforward supply and demand equilibrium—and as a result we devote much of our educational resources towards developing human capital with these abilities.

But as AI progresses, Moravec's Paradox, from an economic viewpoint, makes current education policy look more and more obsolete. Instead of exploiting our comparative learning advantage, we continue to compete against machines in mastering knowledge-based recall and reasoning, which is a struggle that we cannot win. As a result, we are graduating students with knowledgebases/skillsets that have a T_r fast approaching zero. In other words, the job tasks requiring the knowledgebase/skillsets they went to school to study, will be automated before the students can ever monetize what they learned. Already the consequences of this education failure are contributing to a mismatch in the labor market. There are many people looking for work, while at the same time many jobs are going unfilled. This is in part because students are taught how to do jobs that machines perform better than humans, while not being taught jobs that machines perform worse than humans. It is clear from these considerations that the purpose of education needs to be rethought. While there are multiple reasons for this labor mismatch—demographic shifts, disruptions from climate change, immigration barriers, political dysfunction—education is an important factor. In addition, effective education is necessary for solving the other problems.

6.1   Human Learning and Performance

In this book, learning is separated into two dimensions—expertise and interpersonal—with the difference being those tasks requiring expertise have stable end points, while those requiring interpersonal do not. In general, jobs require a mix of both kinds of tasks, but some jobs have tasks primarily along the expertise dimension, while other jobs are composed of tasks that are primarily interpersonal. Humans as they develop and progress through life must acquire knowledgebases/skillsets along both dimensions and must learn much more than just how to perform tasks that can be monetized in the labor market. Life outside of a paying job requires moving through society by navigating a complex web of interpersonal relationships and technologies. While formal education focusses on the expertise dimension, students also learn along the interpersonal dimension while attending school. Indeed, the act of going to school requires knowledgebases/skillsets along both dimensions.

The underlying mechanisms for human learning and performance, however, are very different than for machines, even when the tasks are the same. Both humans and machines can use formal analysis and computation in performing a job task. But for humans, in comparison to machines, these processes are extremely slow and are constrained by negligible amounts of short-term working memory. What humans can perform rapidly and efficiently is pattern recognition—a task that for machines is extremely demanding on computational resources.

Artificial intelligence experiments using the game of chess illustrate these differences. The job is the same—playing chess—but the learning and performing methods that humans and machines employ are vastly different. Chess has a stable endpoint—checkmate—and can be repeated, so learning chess is mostly movement in time along an expertise dimension. Human chess players employ knowledge-based recall, reasoning, and computation, but have only a tiny fraction of the memory, accuracy, and speed that machines have when playing chess.

However, it took more than 20 years from the earliest appearance of commercially available chess computers in the 1970's to the first defeat of world champion in 1997 by a computer specially designed for the sole purpose of beating the world champion—it was not a general-purpose computer (Hsu 2002). Somewhat counterintuitively, chess is a hard problem for machines despite their enormous advantages in memory recall, and computational speed and accuracy. For decades, human chess experts could perform at a level much better than machines by using mostly pattern recognition augmented with relatively little, in comparison to machines, computational processing.

That human experts rely extensively on pattern recognition in playing chess is seen in the common sparing practice among chess players of limited-time chess—so-called "speed" or "blitz" chess—in which the two players are restricted to 5 minutes each in making all their moves in a game, which is about 10 seconds per move on average. Clearly under these sped-up conditions, that severely limit the use of computational processes, the players must rely primarily on pattern recognition in move selection. Yet, the quality of play in these kinds of games and their outcomes are reasonably commensurate with the level of expertise the players possess (Calderwood et al. 1988).

Even more counterintuitive are AI experiments involving the game of poker. It took an additional 20 years after the World Chess Champion was defeated by a machine, for researchers to create a machine that could play poker at a level comparable to top professional poker players (Moravcik 2017). On the surface poker appears to be a much simpler game than chess, but in fact, it is a much more complicated—especially from an AI perspective.

Like chess experts, poker experts rely heavily on pattern recognition augmented by reasoning and computation. However, poker has the complication of incomplete information, so calculations yield statistical answers, not certainties like in chess which, in contrast to poker, is a game of complete information—nothing is hidden in a chess game. In addition, poker requires interpersonal dimension learning in a way that chess does not. A chess player can ignore their opponents' behaviors and focus only on their moves. A poker player must study their opponents' behaviors, build mental models of each opponent's decision-making process, account for the possibility that an opponent might take actions that intentionally deceive, and consider how each opponent is mentally modeling the behavior of the other players, including his or herself. As if this isn't complicated enough, all this behavioral modeling is evolving in real time as the game is being played and the poker player, along with all the other players are also adjusting. Unlike chess, which is always played one-on-one, poker is played with any number of between 2 and 10 players at the same time, and the number of players participating also affects the strategies in use.

In contrast to chess, learning poker is movement along the interpersonal and expertise dimensions at the same time because the job of playing poker requires both kinds of tasks. Unlike chess, poker does not have a stable endpoint. Nothing equivalent to checkmate exists in poker. Poker players measure performance in terms of rates—hourly win rate for cash game players; return on investment for tournament players. Even the quality of the decisions made cannot be expressed in terms of outcomes, but rather a statistical concept called expected value.[1] Ranking poker players using a distribution function, as is done in chess, is not done in poker. The absence of a stable end point and the statistical uncertainties in outcomes because of incomplete information, make the use of a distribution function problematic. However, over the long run some poker players clearly outperform others, which is evidence that consistent success at poker involves a high degree of skill (Dedonno and Detterman 2008). In fact, poker would not be such a difficult AI problem if success didn't require a high degree of skill (Billings et al. 2002).

In Carses' language, chess is a finite game, while poker is an infinite game. In the language of this book, chess is a Judgment Game while poker is a Learning Game. Poker is a kind of Learning Game because the decision inputs have time-dependencies. A chess grandmaster can look at a given position, frozen in time, and state what move is best. The game is about judgment at an instant in time. A poker professional cannot state the best action for a given situation, because the best action is usually context dependent, and often depends on the events that occurred before the situation arouse and events expected to occur afterwards. The game is about real-time learning.

Jobs, like poker, that require a significant number of interpersonal tasks in addition to expertise, are more like Learning Games, while jobs, like chess, that depend mostly on expertise are more like Judgment Games. Considerations in this book suggest that

---

[1] Expected value is the amount of money that a player makes on average from a specific action in a poker game. Expected values can be positive (money-making over repeated plays) or negative (money-losing over repeated plays). It is the expected value of a play that determines whether it is a good or a bad decision. Because of the statistical nature of poker, specific outcomes are unpredictable, while long-term averages are predictable. Therefore, successful poker players learn to ignore specific outcomes and make decisions based on averages.

in the future humans will find their comparative learning advantage in jobs that are Learning Games, not jobs that are Judgement Games. The expertise required for a Judgment Game will always have a T_c that for humans will be biologically limited and cannot be significantly sped up. Machine T_c's for expertise have no such limitation and in principle could be sped up many orders of magnitude as technology advances.

6.2 Conclusion

The first wave of the digital revolution brought computers that were tools for enhancing worker productivity. Tasks such as word-processing, data analysis, record keeping, research, communication, person-to-person collaboration all existed before the advent of computers. However, the introduction of the computer, followed by the networking of computers, resulted in a boon in productivity as workers could perform these activities at a higher volume and faster rate.

The next wave in the digital revolution, in which machine learning is incorporated into the workplace, will fundamentally alter the relationship between machines and workers. Machines will no longer be just productivity enhancement tools like computers, but in a sense, co-workers because a division of labor will need to be negotiated between the artificial intelligence and the human intelligence. It is also becoming clear that economic forces will drive the development of machine learning toward the creation of specialized super-intelligences rather than the replication of general human intelligence. The arguments in this book suggest that in dividing labor, humans should focus on jobs that can be modeled as Learning Games and machines on jobs that can be modeled as Judgment Games.

Traditionally a person's identity has been based on skillset/knowledgebase capacity, in other words what a person does—plumbing, surgery, retail sales, and so on. People do not identify by what they are capable of learning—in fact people often do not know their learning capabilities. In addition, interpersonal skills are critical to success, but not always explicitly acknowledged. However, the modern economy is transitioning from a past that valued knowing/performing to a future that emphasizes learning. Education is not keeping up with this change because of its focus on knowing/performing over learning. In addition, education focusses mostly along the expertise dimension of learning, with the interpersonal dimension acknowledged, but is not often an intentional learning outcome.

Technological and culture change has always been part of the human condition, which has meant that the future has always been more about learning than knowing/performing than people have been willing to explicitly acknowledge. The advent of the industrial revolution accelerated this trend because machines could be built to replace human performance. To stay ahead of the machines, humans had to learn over the course of their lifetimes. But now, machines can learn. Humans must compete not only against machine performance, but with machine learning. The rapidly increased rates of technological and cultural change make the ability to learn even more important in the future.

Therefore, the analysis in this book leads to the following counterintuitive recommendations for education and labor policies going forward:

Education

1. *The current trend of emphasizing technical training over a liberal arts education in the arts, sciences and humanities is completely backwards.* Technical knowledgebases/skillsets have almost by definition, relevant times (T_r's) much less than an expected human lifetime. In the future it can also be expected that characteristic learning times (T_c's) for machines regarding technical knowledgebases/skillsets will be much less than the T_c's for humans. Therefore, machines will have a significant comparative learning advantage for technical knowledgebases/skillsets in comparison to humans. Given this inevitability, a liberal arts education—that includes math and science (these are liberal arts)—is the only kind of education that makes any sense.

2. *The articulation of "college and career" readiness knowledgebases/skillsets is an exercise in futility that consumes valuable time and resources for no productive end.* Because T_r's for jobs are becoming so short, a detailed list of knowledgebases/skillsets to teach students is guaranteed to be obsolete before implementation, let alone before any student embarks on a career. More important than an articulated knowledgebase/skillset are habits of the mind such as critical thinking, problem solving, communicating, adapting, which can only be developed through practice, experience, and mentorship.

3. *Learning along the interpersonal dimension should be an intentional outcome of education, not a byproduct that may or may not occur.* Traditionally, education has focused on teaching *individual* expertise. The word "individual" is emphasized because grades are assigned to individuals, often using a competitive scoring system, and in many circumstances, collaboration is considered a form of cheating. However, it is known that interpersonal knowledgebases/skillsets are of great value in the modern workplace because most work is collaborative. Education systems should recognize this reality and intentionally prepare students for it.

Labor

1. *Employers should screen candidates for learning ability rather than specific skillsets.* Currently there is a mismatch in the labor market with employers unable to find qualified employees for open jobs and employment seekers unable to find jobs for which they qualify. This mismatch arises because employers expect employees to have knowledge and experience for new jobs before the jobs have been created. This impossibility results in frustration for both buyers and sellers in the labor market. To resolve this issue, employers need to be willing to pay for on-the-job learning and hire people with demonstrated success in learning. This would of course a necessitate a change in employer thinking. They would need to see employees as long-term investments in human capital rather than as interchangeable disposable parts in a corporate machine.

2. *Employees should expect that their current knowledgebases/skillsets will be of temporary value in the labor market.* The idea of going to school early in life to acquire a specific knowledgebase/skillset that can then be monetized over an entire "career" is obsolete. Learning must be a lifelong pursuit and the singular use of the word "career" needs to be changed to the plural "careers." Education early in life should focus on the meta-skills needed for learning—learning how to learn—rather than a job-specific knowledgebase/skillset.

3. *Government should be systematic in aiding worker job transitions.* Worker dislocation is often treated as an aberrant economic event. Government aid for retraining workers with obsolete knowledgebases/skillsets is often patchwork and usually in response to unrest. This is in stark contrast to early-life education—K-12 school and college—which if not publicly funded, is often publicly subsidized. Just as it is considered normal for all young people to go to school, it should be considered normal for all adults to move in an out of education programs throughout their lives. There should be programs and publicly funded subsidies for life-long learning. The expenses should not be shouldered entirely by the dislocated workers.

In addition to these recommendations for labor and education, this book also provides a framework for AI researchers as they develop machine learning to an extent that AI evolves into the role of co-worker rather than just a productivity enhancement tool. If AI is to work *with* humans, it must have knowledgebases/skillsets along the expertise *and* interpersonal dimensions. The framework constructed in this book provides a practical means for separating the interpersonal dimension from the expertise dimension. Interpersonal and expertise knowledgebases/skillsets for machines can be tested separately using static Judgment Games and/or trained separately using time-dependent Learning Games. If, in the future, machines simulate human-level cognition they must be able to play the role of the Interrogator in the Imitation Game, not just substitute for Player A.

Moreover, it is the understanding of learning that needs to change. Learning, which is time-dependent, is not the same as knowing, which is static. In our economic system, learning is an expense, while knowing generates revenue. As a result, value is placed on knowing, while learning is regarded as a profit-eroding cost to control. Of course, there can be no "knowing" without "learning," which is why education is often referred to as an "investment." But, as machines get better at learning, humans will need to broaden the concept of "learning" beyond the economic. If machines can learn to do our work for us, what will be our motivation to learn? Under the current human condition, a world with no human work is one without meaning and as a result, humans lack motivation to learn.

When Adam was expelled from the Garden of Eden, God cursed him with work. "Cursed is the ground because of you; through painful toil you will eat food from it all the days of your life." (Genesis 3:17, *Bible*, New International Version, 1978). Paradoxically, despite the curse, humans are terrified of a return to Eden. A world in which the machines that we have created do all the work, leaving us with only leisure activities, is viewed as a dystopia. The resolution of this paradox will require a rethinking not only of our economic systems, but also of how humans *learn* to derive meaning in their lives.

References

Billings D, Davidson A, Schaeffer J, Szafron D (2002) The challenge of poker. Artif Intell 134(1–2):201–240. https://doi.org/10.1016/S0004-3702(01)00130-8

Calderwood R, Klein GA, Crandall BW (1988) Time pressure, skill, and move quality in chess. Am J Psychol 101(7):481–495. https://doi.org/10.1111/j.0956-7976.2004.00699.x

Dedonno MA, Detterman DK (2008) Poker is skill. Gaming Law Rev 12(1):31–36. https://doi.org/10.1089/glr.2008.12105

Genesis 3:17. Bible, New International Version (1978)

Hsu FH (2002) Behind deep blue: building the computer that defeated the world chess champion. Princeton University Press

Moravcik M, Schmid M, Burch N, Lisy V, Morrill D, Bard N, Davis T, Waugh K, Johanson M, Bowling M (2017) DeepStack: expert-level artificial intelligence in no-limit poker. Science 356(6337):508–513. https://doi.org/10.1126/science.aam6960

Moravec H (1988) Mind children: the future of robot and human intelligence. Harvard University Press

Pinker S (1994) The language instinct: how the mind creates language. William Morrow and Company